AFRICAN ISSUES

The Lie of the Land

Challenging Received Wisdom on the African Environment

AFRICAN ISSUES

Series Editor Alex de Waal

Fighting   War, Youth
for the   & Resources
Rain Forest   in Sierra Leone

PAUL RICHARDS

The Lie   Challenging
of the   Received Wisdom on the
Land   African Environment

Edited by
MELISSA LEACH & ROBIN MEARNS

Peace   How the IMF
without   Blocks Rebuilding
Profit   in Mozambique

JOSEPH HANLON

The Lie of the Land

Challenging Received Wisdom on the African Environment

Edited by
MELISSA LEACH
& ROBIN MEARNS
Institute of Development Studies
University of Sussex

The International
African Institute

in association with
JAMES CURREY
Oxford
HEINEMANN
Portsmouth (N.H.)

The International African Institute
London
in association with
James Currey Ltd
73 Botley Road
Oxford OX2 0BS

Heinemann
A division of Reed Elsevier Inc
361 Hanover Street, Portsmouth,
New Hampshire 03801-3912

British Library Cataloguing in Publication Data
The lie of the land: challenging received wisdom on the
African environment. – (African issues)
1. Environmental policy – Africa 2. Environmental degradation
- Africa
I. Leach, Melissa II. Mearns, Robin III. International African Institute
333.7'096

ISBN 0-85255-409-5 (James Currey Paper)
ISBN 0-85255-410-9 (James Currey Cloth)

Library of Congress Cataloging-in-Publication Data applied for

ISBN 0-435-07408-3 (Heinemann Paper)
ISBN 0-435-07407-5 (Heinemann Cloth)

Typeset by
Exe Valley Dataset
Exeter, Devon
in 9/11 Melior with Optima display

Printed in Great Britain by
Villiers Publications
London N3

CONTENTS

LIST OF FIGURES & TABLES

i

Figures

Tables

ACKNOWLEDGEMENTS

This book originated in a conference held at the Institute of Development Studies (IDS), University of Sussex, on 13–14 September 1994. Entitled 'Escaping orthodoxy: environmental change assessments, local natural resource management, and policy processes in Africa', the meeting was planned to bring together the majority of the cases included here, and to discuss their wider implications. The conference was funded jointly by IDS and the UK Economic and Social Research Council (ESRC), as the last in a series of four ESRC workshops on African environmental issues initiated by Phil O'Keefe. We would like to thank Ann Watson for her efficiency and good humour as conference administrator, the staff of the IDS Teaching Area for their practical support, and Wendy Kenyon for being conference rapporteur. Among the conference participants, and in addition to the book contributors, thanks are due to the session discussants, Mike Arnold, Allan Hoben, Cecile Jackson and Richard Moorehead; to paper-givers Richard Grove, Michael Mortimore and Paul Richards; and to Sam Bickersteth, Robert Chambers, Kate Wellard and Brian Wynne for their invited reflections on broader research and policy perspectives. The introduction to this book owes a great deal to the contributions of these individuals and to points raised in general discussion, although responsibility for the ways we have represented them here rests with us alone.

In preparing the book, we are very grateful for the administrative and secretarial help of Kim Collins, administrative assistant to the IDS Environment Group. We would particularly like to thank James Fairhead, Ann Hudock and Ian Scoones for their valuable comments and support in the preparation of the introduction. Chapter Eleven is a revised version of an article by Allan Hoben which originally appeared in *World Development*; it is published here with the permission of Elsevier Science Ltd. Chapter Three appears in a different form in G. Chapman and T. Driver (eds) *Timescales of Environmental Change* (Routledge, 1996). And for the title, we must thank Alex de Waal.

Melissa Leach and Robin Mearns
IDS

W. M. Adams is a lecturer in Geography at the University of Cambridge. He has carried out extensive field research on environment and development issues in Africa and is the author of several books including *Nature's Place: conservation sites and countryside change* (Allen and Unwin, 1986), *Green Development: environment and sustainability in the Third World* (Routledge, 1990) and *Wasting the Rain: rivers, people and planning in Africa* (Earthscan, 1992).

William Beinart is Professor in the Department of Historical Studies, University of Bristol. His recent publications include *Twentieth-Century South Africa* (Oxford University Press, 1994) and with Peter Coates, *Environment and History: the Taming of Nature in the USA and South Africa* (Routledge, 1995). He is currently working on land reform issues in South Africa and on an environmental history of the stock-farming areas of the country.

Daniel Brockington is carrying out research in the Anthropology Department at University College London. He has field experience in South Africa and in Tanzania and is currently working on the interactions of wildlife conservation and community development in pastoralist groups around Mkomazi Game Reserve, as well as on the possibilities for community participation in natural resource management.

Reginald Cline-Cole is the geographer in the multi-disciplinary Centre of West African Studies at the University of Birmingham. He has field experience in Sierra Leone, Nigeria, Burkina Faso and Ethiopia. His research has concentrated on rural energy, forestry-related issues, environmental history, society–nature relations and land. His publications include *Woodfuel in Kano* (United Nations University Press, 1990) and articles in *Africa, World Development, Land Use Policy,* and *Forest and Conservation History*.

James Fairhead is a social anthropologist and currently departmental lecturer in anthropology and development at the University of Oxford, Queen Elizabeth House International Development Centre. He has field experience in Central and West Africa, and has carried out research on the social and technical aspects of agro-ecological knowledge; agrarian and environmental history; forest and forest-savanna ecological dynamics, and the theoretical and methodological context of assessments of West African vegetation change. Recent publications include *Misreading the African Landscape: society and ecology in a forest-savanna mosaic* (with Melissa Leach, Cambridge University Press, 1996) and articles in *Africa, African Affairs, Agriculture and Human Values* and *World Development*.

Allan Hoben is Professor of Anthropology and Fellow of the African Studies Center at Boston University, Boston, MA. He has extensive field research experience on land tenure and agrarian development in Ethiopia. He has also done research, consulting and advising on problems of agricultural and rural development in many other African nations and is currently carrying out a study of environmental policy in Africa. His publications include *Land Tenure Among the Amhara of Ethiopia* (University of Chicago Press, 1972), articles in academic journals, and numerous research reports.

Katherine Homewood is an anthropologist with a biological background. She is Reader in Anthropology at University College London. She has field experience in Kenya, Tanzania, Algeria and Burkino Faso. Her research has focused on the interactions of conservation and development in sub-Saharan Africa, with particular focus on the implications of changing land use patterns for environmental values and human welfare. Her publications include *Maasailand Ecology: wildlife conservation and pastoralist development in Ngorongoro Conservation Areas, Tanzania* (with W. A. Rodgers, Cambridge University Press, 1991), together with articles in *Journal of Agricultural Science; Journal of Applied Ecology; Ecology of Food and Nutrition*, and *Africa*.

Melissa Leach is a social anthropologist and Fellow of the Institute of Development Studies at the University of Sussex. She has extensive field experience in West Africa, especially Sierra Leone and the Republic of Guinea. Her research has focused on environmental history; forest and forest-savanna ecological dynamics; gender–environment relations; agrarian social change and the social construction of environmental knowledge. Recent publications include *Rainforest Relations: gender and resource use among the Mende of Gola, Sierra Leone* (Edinburgh University Press and Smithsonian Institution, 1994), *Misreading the African Landscape: society and ecology in a forest-*

savanna mosaic (with James Fairhead, Cambridge University Press, 1996) and articles in *World Development, Africa, African Affairs* and *Environment and History.*

Robin Mearns is a geographer and Fellow of the Institute of Development Studies at the University of Sussex. He has field experience in Papua New Guinea, East and Horn of Africa, Mongolia and post-Soviet Central Asia, and has carried out research, training, operational and advisory work on land and agrarian reform, pastoral institutions, common property resource management, household energy, rural development forestry, and poverty–environment interactions. His publications include *Beyond the Woodfuel Crisis* (Earthscan, 1989) and articles in *Journal of Development Studies, Cahiers des Sciences Humaines, Development and Change* and *Nomadic Peoples.*

Ian Scoones is an agricultural ecologist and Fellow of the Institute of Development Studies at the University of Sussex. He has field experience in Southern and East Africa, and has carried out research on issues of natural resource management in dryland areas, with a recent focus on pastoral issues. Recent publications include *Beyond Farmer First: rural people's knowledge, agricultural research and extension practice* (ed. with J. Thompson, IT Publications, 1994); *Living with Uncertainty: new directions in pastoral development in Africa* (ed. IT Publications, 1995), and *Hazards and Opportunities: farming livelihoods in dryland Africa – lessons from Zimbabwe* (with C. Chibudu *et al.*, Zed Books, 1996).

Michael Stocking is a tropical soil scientist and Professor of Natural Resource Development at the University of East Anglia. With field experience in sub-Saharan Africa, South America and South and Southeast Asia, his work involves erosion monitoring, soils investigations and the relationship between land degradation and vegetation productivity. Agroforestry, intercropping, multiple land use and small farm development feature in his work, along with biodiversity and integrated conservation-development projects. He is an adviser to FAO, UNEP, WWF and the United Nations University, as well as national agencies. His publications include papers in *Global Environmental Change, Ambio,* and *Catena,* and an edited book of readings on topical environment–development issues, *People and Environment* (UCL Press, 1995).

Jeremy Swift is a development economist and Fellow of the Institute of Development Studies at the University of Sussex, with experience in East and West Africa, Southwest and Central Asia. His research focuses on extensive livestock economies in arid and mountain ecosystems, including especially land tenure, local institutions and food security. He has written extensively on these issues and on famine.

Mary Tiffen is a socio-economist and historian, formerly of the Overseas Development Institute, London. She has carried out much research and consultancy in sub-Saharan Africa and the Middle East on the processes of agricultural development, with particular reference in the last fifteen years to population–environment relations, soil conservation and irrigation. She has been particularly interested in the dynamics of change over time, and the implications this has for policy. Her books include *The Enterprising Peasant* (HMSO, 1976); *Theory and Practice in Plantation Agriculture: an economic review* (with Michael Mortimore, ODI, 1990) and *More People, Less Erosion: environmental recovery in Kenya* (with Michael Mortimore and Francis Gichuki, Wiley, 1994). Recent articles have been for *World Development, Development and Change, Journal of International Development,* and *Recherche Développement.*

1

Environmental Change & Policy

MELISSA LEACH
& ROBIN MEARNS

Challenging Received Wisdom in Africa

Introduction

The driving force behind much environmental policy in Africa is a set of powerful, widely perceived images of environmental change. They include overgrazing and the 'desertification' of drylands, the widespread existence of a 'woodfuel crisis', the rapid and recent removal of once-pristine forests, soil erosion, and the mining of natural resources caused by rapidly growing populations. So self-evident do these phenomena appear that their prevalence is generally regarded as common knowledge among development professionals in African governments, international donor agencies, and non-governmental organizations. They have acquired the status of conventional wisdom: an integral part of the lexicon of development. Yet as shown by accumulating research assembled in this book, these images may be deeply misleading.

The resurgence of concern over the global environment in international development during the late 1980s and early 1990s has given such images a new and vigorous lease of life. *Agenda 21*, the global plan of action adopted at the 1992 Earth Summit in Rio de Janeiro, claims that desertification affects 'about one sixth of the world's population, 70 per cent of all drylands, amounting to 3.6 billion hectares, and one quarter of the total land area of the world' (UN, 1992: 98). A result is 'widespread poverty' (*ibid*.: 98). A prestigious article published in *Science* claims that '[a]bout 80 per cent of the world's agricultural land suffers moderate to severe erosion' (Pimentel *et al.*, 1995: 1117), leading to the destruction and abandonment of arable land attributable directly to nonsustainable farming practices; and that '[m]ore than half of the world's pasturelands are overgrazed and subject to erosive degradation' (*ibid*.: 1117). And research carried out by the World Bank in western and central Africa finds that under conditions of rapid population growth, '[t]raditional farming and livestock husbandry practices, traditional dependency on wood for energy and for building materials, traditional land tenure arrangements, and traditional burdens on women . . . became the major source of forest destruction and degradation of the rural environment' (Cleaver, 1992: 67).

Such views of environmental change in Africa are not restricted to professional circles. They are also popularized in the news and current affairs media in developed and developing countries alike, and help to build support among the general public for the field operations of charitable organizations, designed to halt the forces of environmental destruction. Images of starving children, and the attribution of blame to natural environmental causes, have become an integral part of the way Africa is perceived in the North. They are signposts to the lie of the land: the reasoning behind them is taken for granted and rarely questioned.

These orthodoxies assign to Africa's farmers, hunters and herders a particular role as agents, as well as victims, of environmental change. If current trends are to be reversed, it is implied, local land-use practices will have to be transformed and made less destructive. Yet the development policies and programmes that result commonly prove to be at best neutral and at worst deleterious in their consequences for rural people and for the natural-resource base on which their livelihoods often substantially depend.

Take the notion of the 'woodfuel crisis' as an example (G. Leach and Mearns, 1988; Mearns, 1995). Felling trees for firewood and charcoal is often assumed to be a prime cause of deforestation in Africa, which thus presents a classic case of demand for fuel outstripping supply. It has been conventional to assess the scale of the problem by comparing current woodfuel consumption with current stocks and annual growth of trees. Typically, this comparison identifies a shortfall, which is assumed to be made up by depleting the standing stock of trees. The supply 'gap' is then projected into the future, often in direct proportion to population growth, so that it widens even faster as sustainable yields diminish. There is a crisis. The logical solution is also implicit from the starting assumptions: namely, to plant trees on a colossal scale to close the woodfuel supply gap, and to introduce more efficient cooking stoves to reduce demand. As the 'woodfuel crisis' is or will be so severe, so the argument goes, trees must be planted on a scale far in excess of the capacity of rural people themselves to respond. Government forestry departments, with financial and technical support from aid agencies and non-governmental organizations, must intervene.

By re-framing the problem, however, many authors have now shown such solutions to be inappropriate. They point out that even when projections are made more 'realistic' by correction to allow for factors such as tree regrowth after cutting, as long as the basic assumptions remain intact, three important facts are overlooked. One is that most woodfuel in sub-Saharan Africa comes not from cutting live trees for fuel, but from surplus wood left over from clearing land for agriculture, or from lopping branches off trees standing on farms that are valued for many other purposes besides fuel supply (e.g. shade, fruit, or building poles). Second, wood supply as conventionally defined by foresters greatly underestimates the woody biomass resources available to people as fuel, for example from smaller trees, bushes and shrubs. Third, where people really do perceive a physical shortage of woodfuel, or a need for trees for other reasons, they tend to respond in various ways, whether to reduce fuel consumption, or to plant or encourage the natural regeneration of trees.

Seen from this broader perspective, there is not just one, very big problem of energy supply, but many smaller problems of command over trees and their products to meet a wide range of basic needs, including food, shelter, income and investment. Since these problems take different forms for different people in different places, the range of potential solutions is equally diverse. Rather than simply planting trees for their own sake, the focus shifts to intensifying agricultural production so that output per unit area is increased, but in ways that enable people better to integrate trees with crop production.

The way in which problem and solution are framed in the case of the woodfuel crisis offers a classic example of how 'received wisdom' about environmental change obscures a plurality of other possible views, and often leads to misguided or even fundamentally flawed development policy in Africa. A sustained challenge to received wisdom about the woodfuel crisis has been built up through careful research, but this case is far from unique. Similar challenges to many other orthodoxies concerning African environmental change have been mounted over the last decade or so. This book brings much of this work together for the first time. The ten cases assembled here illustrate challenges to received wisdom across a range of major concerns in contemporary environment and development debates, including desertification, deforestation, rangeland degradation, soil erosion, and the depletion of wildlife biodiversity. Based on detailed, historically informed research, the cases are drawn from a range of African environments, from the drylands of the Sahel and southern Africa, to the humid tropical forests of West Africa. While the contributors are variously anthropologists, historians, ecologists, geographers and economists, they share a sensitivity to the gains to be had from interdisciplinary approaches to examining processes of environmental change. They are also among those best-known for challenging orthodoxies in their respective fields, having generally played prominent roles and published widely in the debates addressed here.

Many of the arguments put forward in the chapters will be familiar, then, although in several cases they are supported by entirely new material. Where this book is pioneering is in drawing these cases together so that their commonalities may be explored. While the debates represented here have been building for some time, challenges to received wisdom have now, by the mid-1990s, reached a critical mass which for the first time permits comparative analysis. Such an exercise is, we argue, not only timely but of critical importance, because as this introduction explores, the insights to be gained from reflecting on these cases together add up to far more than the sum of their parts.

One common denominator of the received ideas considered here is that they rest on neo-Malthusian assumptions concerning the relation-ships between society and environmental change. The symbolism of neo-Malthusian images is deeply embedded in Western popular culture and religion (Hoben, 1995). With the advent of space travel in the late 1960s, the newly projected image of the earth from space gave rise to the icon of 'spaceship earth', which so emphatically conveyed the impression of a fixed natural-resource base and inspired the 1970s environmental movement. But as Grove (1995) shows, pre-Sputnik, 'islands' and even 'gardens' provided equally potent 'global analogues ... of society, of the world, of climate, of economy' (*ibid.*: 13) which helped to shape European attitudes towards nature.

Some of the case studies here demonstrate that orthodox evaluations of environmental change, and hence of local land-use practices, are demonstrably false. The land – or rather, certain representations of it – can indeed lie. Other cases do not fundamentally challenge received wisdom, but do show how it has exaggerated the magnitude of problems such as forest or soil loss. We want to ask why, if the received wisdom can be shown to be misguided or just plain wrong, is it so pervasive and resistant to change? In this introduction, we argue that the answers to this question are to be found in the sociology of science and in the sociology of development; in other words, in the broader historical, political and institutional context for science and policy. An inextricable part of this nexus – and the reason why it is so important – are the profound practical implications for farmers, herders, and other land users in Africa. The chapters consider how received ideas about environmental change have been translated into particular forms of policy intervention, and with what practical consequences. Where appropriate, the cases try to indicate alternative approaches to the way problems are framed and solutions formulated, which frequently call for radical reassessment of the way Africa's farmers and herders are implicated in processes of environmental change.

Contradictory evidence from a single case study cannot, of course, entirely refute an orthodoxy. The fact that forest loss or range degradation may have been misinterpreted in one place does not necessarily mean they are being misinterpreted everywhere. But the fact that contradictory cases exist certainly casts doubt over the general applicability of received views, and calls for a more critical examination of the evidence that supports other cases, in case they too prove to counter dominant opinion. The received wisdom under scrutiny in this book, then, is the product of particular interpretations of the relationship between environmental change and people's behaviour. This collection of case studies exposes such interpretations as selective and open to challenge.

Challenging received wisdom

The exposure of the lie in the land encapsulated in this book comes from three principal angles: history, ecology, and social anthropology, as well as recent fruitful cross-fertilization between them in the analysis of African environments.

A number of popular myths about environmental change have their origins in attempts to infer process from form; that is, to make assumptions about the history of a given landscape on the basis of a 'snapshot' view of its current state or on data gathered over a few years at most. Several case studies here, however, demonstrate the importance of using historical and 'time-series' data sets of various types to study processes of landscape change more directly; documenting history rather than inferring it (cf. Fairhead and Leach, 1996).

The science of ecology itself has helped turn attention to time-series analysis by questioning the validity of baselines such as a 'climax vegetation community', or static concepts such as 'carrying capacity' – notions which have hitherto been integral to the scientific validation of orthodox views of environmental change. Older theories are now yielding to a greater pluralism in ecological thinking (cf. McIntosh, 1987).

Attention to historical detail, and the shedding of theoretical straitjackets in ecology, have converged with a better understanding of the land-use practices of Africa's farmers and herders, and of their own ecological knowledge and views of landscape change. This is amply provided in recent work by social anthropologists and others (e.g. Croll and Parkin, 1992; Richards, 1985; Ellis and Swift, 1988; Fairhead, 1992; Chambers *et al.*, 1989, Warren *et al.*, 1995), following some notable precedents (e.g. Allan, 1965; De Schlippe, 1956). Indeed, the application of recent approaches in ecological history often reveals the logic and rationality of 'indigenous' knowledge and organization in natural resource management. By contrast, received wisdom would have much of the blame for the vegetation change perceived by outsiders as environmental 'degradation' rest with local land-use practices, whether labelling them as ignorant and indiscriminate or – more commonly – as ill-adapted to contemporary socio-economic and demographic pressures. In such accounts, rural people's ecological knowledge is notable mostly by its absence, silenced before it is investigated.

But the reasons why received wisdom has proved so resistant to change have to do with much more than simply 'getting the facts wrong', or 'ignorance' on the part of outside observers of African environments and land users. Particular readings of history and of African land use, particular notions within ecology, and so on, have attractions beyond their own claims to 'truth value'. Their claims to 'truth' may rest on the application of particular methods and theories. They may be shown to serve the purposes of particular institutions, or political or economic interest groups; or to appear logical given the cultural backgrounds of their proponents. But as the cases in this collection demonstrate, such 'truths' may serve to obscure quite different readings of environmental and land-use history. The sustained critique running through these contributions clearly denies the value-neutrality both of the methods employed in the study of environmental change, and of the conclusions derived from them. Before we are able to judge between alternative conceptions of environmental and land-use change, then, we need to be able to specify the conditions of their production.

Bringing together a range of cases therefore allows this book to address three central questions. First, how does received wisdom about environmental change in Africa become established, get reproduced, and in some cases persist even in the face of strong counter-evidence? Second, how is it put to use and with what outcomes? And third, what alternative approaches for policy and applied research are suggested by countervailing views? In the remainder of this introduction, we begin to explore some partial answers to these questions; but first it is necessary to clarify what is meant by received wisdom, by outlining several approaches to theorizing the production of knowledge in public policy.

Theorizing 'received wisdom' in development policy

At the most general level, all the contributors treat received wisdom as an idea or a set of ideas held to be 'correct' by social consensus, or 'the establishment'. Within this broad characterization, there is scope both for more 'structural' explanations, which in emphasizing how social structures produce knowledge, may appear in the extreme like conspiracy theories, and for those emphasizing human agency, which at their own extreme may appear excessively voluntaristic. The kinds of explanation sought in the contributions to this volume inevitably vary in emphasis, given the predilections of individual authors, differences in their case material, and differences in the tenor of the debates their work has engaged with over the years, but most nevertheless tend to combine structural and actor-oriented forms of explanation.

In his seminal study *Development Projects Observed*, Hirschman (1968) showed that 'effective development policies and programmes (i.e., ones that succeed in mobilising funds, institutions and technology) depend on a set of more or less naive, unproven, simplifying and optimistic assumptions about the problem to be addressed and the approach to be taken' (Hoben, 1996; this volume). Hoben describes this as a 'cultural script for action', without which 'it is difficult for donors and aid recipients to mobilize and coordinate concerted action in the face of the many uncertainties that characterize processes of economic, political and institutional change' (*ibid.*). Simplifying assumptions are thus made as a matter of expediency.

Clay and Schaffer (1984) extend Hirschman's work in the field of development sociology and policy studies by questioning the normal premise of policy and planning activities: 'that there is something to be done. Policies make a difference. Different policies could be chosen. There is room for manoeuvre' (*ibid.*: 1). They argue that the mainstream approach to public policy actually 'reduces the margin of manoeuvre towards alternative and better policies' (*ibid.*: 11), owing to structural factors that 'box in' (Long and Van der Ploeg, 1989) individual policy makers within particular institutional establishments and preconceived agendas.

One of the means by which policy makers 'box themselves in' is through labelling (Wood, 1985), referring particularly to 'the way in which people, conceived as objects of policy, are defined in convenient images' (*ibid.*: 1). Labels are put on 'target groups' as passive objects of policy (e.g. 'the landless', 'sharecroppers', 'women'), rather than active subjects with projects and agendas of their own. The disarming shorthand of labelling constructs a problem in such a way as to prescribe a predetermined solution, and legitimizes the actions of development agencies and other public bodies in intervening to bring about the intended results (cf. Long and Van der Ploeg, 1989). Such classifications are 'represented as having universal legitimacy, as though they were in fact natural' (Wood, 1985: 9). Wood argues further that 'labels misrepresent or more deliberately falsify the situation and role of the labelled. In that sense, labels . . . in effect reveal [the] relationship of power between the giver and the bearer of a label' (*ibid.*: 11).

In a similar vein, Roe (1991) has shown how the simplifying assumptions that enable policy makers to act are frequently encoded within 'development narratives'. As a 'story', these have 'a beginning, middle, and end (or premises and conclusions, when cast in the form of an argument) . . . development narratives tell scenarios not so much about what should happen as about what will happen – according to their tellers – if the events or positions are carried out as described' (Roe, 1991: 288).

Nowhere is the power of policy narratives and paradigms illustrated more clearly than in environmental planning in developing countries, as Hoben argues:

The environmental policies promoted by colonial regimes and later by donors in Africa rest on historically grounded, culturally constructed paradigms that at once describe a problem and prescribe its solution. Many of them are rooted in a narrative that tells us how things were in an earlier time when people lived in harmony with nature, how human agency has altered that harmony, and of the calamities that will plague people and nature if dramatic action is not taken soon. (Hoben, 1995: 1,008)

Whatever their truth-value – and as Roe points out, this may be in question – narratives 'are explicitly more programmatic than myths, and have the objective of getting their hearers to believe or do something' (Roe, 1991: 288). By making 'stabilizing' assumptions to facilitate decision-making, narratives serve to standardize, package and label environmental problems so that they appear to be universally applicable and to justify equally standardized, off-the-shelf solutions.

Whether understood in terms of labelling or narratives, what is happening here is the representation of the experiences of those who are seen to be 'the problem' outside their specific historical and geographical contexts. The stabilizing assumptions of policy makers thus substitute for the rich diversity of people's historical interactions with particular environments. Even when they embrace debate, such debates often reduce the world to two dimensions in a simplified and ultimately unhelpful way. Environment and development discourse is replete with examples, frequently taking the form of 'bad'/'good' dichotomies: 'tragedy of the commons' versus common property resource management; farmers' ignorance versus 'indigenous technical knowledge'; Malthusian degradation versus Boserupian intensification, and so on.

Hoben (this volume) highlights more cultural dimensions of this issue, arguing that 'the power of development narratives is enhanced through the incorporation of dominant symbols, ideologies and real or imagined historical experience of their adherents. In this sense they are culturally constructed and reflect the hegemony of Western development discourse' (cf. Hoben, 1995). He suggests that as narratives become influential within environment and development practice, so they help shape their own 'cultural paradigm': namely, specific development programmes, projects, packages and methodologies of data collection and analysis. The 'cultural policy paradigm' thus builds its own foundation, being 'based on concrete exemplars as well as on a set of ideas'.

By received wisdom as applied in African environmental change and policy, then, we mean an idea or set of ideas sustained through labelling, commonly represented in the form of a narrative, and grounded in a specific cultural policy paradigm. It can be understood as a form of 'discourse', in the sense meant by Foucault (1971; 1980) to draw attention to the way in which it embodies relations of power that are constituted through everyday, familiar acts that go unnoticed because they are taken for granted (cf. Milton, 1993). The fact that received wisdom as discourse is embedded in particular institutional structures, active on the ground, not only accounts for its tenacity, but enables it to have real practical consequences, or 'instrument effects' (Foucault, 1979; cited in Ferguson, 1990) that reveal the underlying exercise of power.

Keeping in mind the structure-agency axis referred to earlier, it is important to be explicit about the degree of intentionality at work in producing and reproducing received wisdom and its actual consequences once put into policy. Representations of environmental change and the role of assorted people and organizations in bringing it about are rarely uncontested. Received wisdom should not be conceived of as somehow autonomous, with a life of its own, independent of human agents. Rather, it is at the same time a product of the unintended and intended consequences of the actions of individual human agents, and a part of the structure within which they act and which shapes future possibilities for action (Giddens, 1984; Long and Long, 1992).

In some of the contributions to this volume, the influence of African land users over the way in which environmental change is conceived in the development process appears rather small. In the case of desertification, for instance (Swift, this volume), received wisdom pays little heed to the perceptions of indigenous herders, but rather represents a hegemonic, 'totalising discourse' (Peet and Watts, 1993) in which their position is often relegated to that of resistance to projects imposed. In other cases, however, local land users have had much greater influence over received views of environmental change. There is evidence of considerable interaction between different knowledge systems, which gives rise to new strands of knowledge at their interfaces (Long, 1989; Long and Long, 1992). This certainly seems to have been the case at certain times in parts of the West African forest zone, for example. Fairhead and Leach (this volume) describe how farmers in Guinea have contributed to the State's discourse concerning forest loss when attempting to secure development benefits, for example, despite the very different environmental opinions and experiences they express in other contexts.

The origins and persistence of received wisdom

The received wisdom that we consider in this book is by no means new. In many cases, the ideas that drive contemporary environmental policy in Africa can be traced back to early colonial times. But the reasons for their origins and persistence are to be found at different levels. It might seem reasonable to assert that received wisdoms are held because, at a first approximation, they capture realities important to people's lives and problems. Certainly, their substantive messages, the underlying theories which lend them scientific credibility, and the methodologies from which those messages are derived all merit close attention and critical evaluation in their own right. As the cases will show, however, it is striking that recourse is taken to the same substantive messages, theories and methodologies, time after time, even in situations where they have been shown to hold little validity. To explain this we need to consider the broader social and historical context within which science is used in the service of public policy (Collingridge and Reeve, 1986). Here we consider each of these levels in turn, first examining issues of scientific theory and methodology, and then issues in the sociology of science and of development, and finally the role of popular culture in the construction of environmental meaning. We refer principally to the chapters of this book rather than to other published work on individual cases because each chapter already synthesizes a wider literature on the topic in question, including the contributors' own previous work.

Scientific theory
Only a small number of ideas and theories have been truly pivotal in debates about environmental change and human-environment inter-actions in Africa, but their influence has been enormous. One has been the notion of a 'climax vegetation community': the vegetation that a given climatic zone would support in the absence of disturbance. Another pervasive idea has been the supposed causal link between de-vegetation and declining rainfall. This was strongly evident in nineteenth-century thinking (Grove, 1995), and has been as influential in theories of the derivation of savanna from forest in Africa (Fairhead and Leach, this volume) as in those of desertification further north (Swift, this volume). A third is the idea of carrying capacity: that every set of ecological conditions can support a given number of people and/or livestock which, once exceeded, will lead to a spiral of declining productivity.

Many of these sets of ideas or theories, dominant in ecological science since its inception (Clements, 1916), have their foundation in some notion of equilibrium. This could be the equilibrium between environmental factors (e.g. climate, soils and vegetation) that would prevail in the absence of people, as in the notion of vegetation climax. Or it could be an equilibrium between certain sorts of society and environment (e.g. 'traditional' society in culturally regulated harmony with 'nature', remaining within carrying capacity). In each case, environmental change could be projected as a linear departure from the ideal.

Recent thinking in ecology questions these ideas of equilibrium, instead emphasizing the inherent variability of many ecosystems. In space, assemblages of biotic communities are increasingly thought to resemble more of a patchwork, controlled by edaphic and other abiotic factors often on quite a small scale, than mere variants on some presumed ideal 'climax' community. In time, variability frequently takes the form of state-and-transition dynamics, rendered complex by spatial variations. There is not one unique, ideal state which is 'deflected' into an inferior state upon disturbance. Rather, historically specific conjunctures of conditions may determine unique pathways of transition from one state to another, and may even give the appearance of 'chaotic' fluctuation. This has led the ecologist Robert May to comment on 'the ineluctably contingent nature of such rules and patterns as are to be found governing the organization of communities' (cited in McIntosh, 1987: 322). Sometimes heralded as 'the new ecology' (Botkin, 1990; Worster, 1990a), or ecological pluralism (McIntosh, 1987), such perspectives are certainly opening up new ways of conceptualizing the dynamics of ecological systems whether in drylands (Behnke *et al.*, 1993; Dublin *et al.*, 1990; Scoones, this volume) or forests (Sprugel, 1991; Fairhead and Leach, this volume).

More broadly, the message contained within the scientific theories that underpin received wisdom can be seen to reflect culturally and historically specific representations of 'the environment'. The very concept of an external 'environment', analytically separable from society, can be traced to post-Enlightenment thought in the North (Glacken, 1967; Worster, 1977). Western science rests on the basic assumption that 'natural' phenomena can be investigated separately from human society, except inasmuch as people and their social world are subject to 'nature' and act on 'it'. Such a distinction is, of course, alien to many African societies, in which categories of thought are structured in very different ways and cut across a nature–culture divide (Croll and Parkin, 1992; Fairhead and Leach, 1996; Gottlieb, 1992).

The assumptions of post-Enlightenment science are manifested in several different ways in the received wisdoms explored here. On the one hand, they are evident in views which seek out untouched, pristine nature against which to assess human impact, whether in undisturbed, climax forest vegetation (Fairhead and Leach, this volume); or in the wildlife-rich wilderness of southern and eastern Africa as represented in northern colonial and popular imaginations (Anderson and Grove, 1987; Brockington and Homewood, this volume). Human impact is portrayed in terms of 'anthropogenic disturbance' to an otherwise stable nature.

On the other hand, post-Enlightenment thought is also evident in the view – held from Francis Bacon's time onwards – that society can and should use the technology at its disposal to achieve mastery or dominance over nature so as to satisfy human needs and wants. Others highlight the ideological character of such a view, suggesting that it serves to disguise the real form of domination which is between classes in society (Leiss, 1972). In this volume, Scoones describes how imperatives of administrative control and surveillance underlay the attempts by colonial and post-colonial administrators to impose order and straight lines on the rangeland landscape in southern Africa. Similarly, in northern Nigeria, expatriate foresters have been convinced that only trees planted in lines and orderly plantations constitute a 'proper' use of the drylands (Cline-Cole, this volume). What masquerades as environmental control is often more correctly seen as social control.

Implicit in culturally specific representations of the environment are particular notions of its 'value', as derived from prevailing priorities in natural resource exploitation, or from the biases of particular scientific disciplines. A good example concerns the way professional foresters and ecologists in Africa have conventionally valued closed-canopy or gallery forest – almost defining 'forest' in these terms – so that any conversion of such a vegetation community is seen to constitute 'degradation'. Yet such conversion may be viewed positively by local inhabitants, for whom the resulting bush fallow vegetation provides a greater range of gathered plant products and more productive agricultural land (Davies and Richards, 1991; Leach, 1994). Thus the same landscape changes can be perceived and valued in different ways by different groups; what is 'degraded and degrading' for some may for others be merely transformed or even improved (cf. Beinart, this volume).

Ideas and theories in social science have often converged with those in natural science in the production of received wisdom about environmental problems. Indeed in some cases, social science has provided supporting narratives that reinforce demonstrably false analyses of the nature and causes of environmental change. Hoben (this volume), for instance, describes how a neo-Malthusian narrative concerning the impact of population growth has supported the orthodox view of recent environmental collapse in highland Ethiopia. He counters this with an alternative analysis that suggests there is nothing new in environmental 'flux, crisis and calamity' in that setting, and that it should more properly be attributed to the political economy of the state and its influence over natural-resource use than to demography. While the alternative view does not deny that there are serious problems of soil erosion in highland Ethiopia, it certainly provides a counterweight to the neo-Malthusian narrative that 'exaggerates the rate and magnitude of degradation and misrepresents the role of human agency in causing it' (Hoben, this volume).

Perhaps the best known supporting narrative from social science is the so-called 'tragedy of the commons' argument, used to support received wisdom about dryland environmental change. It runs as follows:

> Picture a pasture open to all. It is to be expected that each herdsman will try to keep as many cattle as possible on the commons. Such an arrangement may work reasonably satisfactorily for centuries because tribal wars, poaching, and disease keep the numbers of both man and

beast well below the carrying capacity of the land. Finally, however, comes the day of reckoning, that is, the day when the long-desired goal of social stability becomes a reality. At that point, the inherent logic of the commons remorselessly generates a tragedy . . . Ruin is the destination toward which all men rush, each pursuing his own best interest in a society that believes in the freedom of the commons. (Hardin, 1968: 1244)

This argument, used as a metaphor by Hardin and others (e.g. Ehrlich and Ehrlich, 1990) to account for population pressure on resources in general, has been critically analysed as a development narrative by Roe (1991; 1994). A strong counter-narrative refutes it, arguing that Hardin confused 'common property' with open access; that in a true commons situation, local institutions facilitate cooperation between users such that resources can be managed sustainably; and that 'tragedies' are a result of the breakdown of such arrangements, for example through state intervention (e.g. Bromley and Cernea, 1989; Bromley, 1992; Ostrom, 1990; Shepherd, 1989). Yet Swift (this volume) shows how the 'tragedy of the commons' has been marshalled consistently to support the conviction that the world's deserts are 'on the march', in spite of an absence of reliable empirical evidence to support that view. He finds the explanation for its persistence in the fact that it serves well the interests of donor agencies and national governments in perpetuating various forms of planned development. The finger of blame has been pointed in a different, but perhaps equally misleading, direction in the case of South Africa, described by Beinart (this volume). Here, rangeland degradation was said to have been caused by the destructive farming practices of white settlers; a view which conveniently suited the political agendas of those opposing the apartheid regime. Yet time-series evidence tells a rather different story, of general stability in grassland composition over the period in question.

The persistence of received wisdoms may also depend on what is left out of the analysis; on the way that dominant ideas serve to highlight certain aspects of local farming and environmental management while excluding others. Tiffen (this volume) gives several examples of such 'blind spots' on the part of outside observers of smallholder agriculture. Researchers and others working in Kenya's Machakos district, for instance, failed to 'see' farmers' investments in soil conservation and farm landscape improvement, because of a conviction that they were 'resource-poor' and therefore must lack the necessary capital to make such investments (cf. Chambers, 1990). Farmers' investments in enriching their local environment through the deliberate manipulation of ecological processes to encourage tree regrowth were also invisible to, or ignored by, outsiders and government officials in Guinea's forest-savanna transition zone (Fairhead and Leach, this volume) and in the Kano close-settled zone of Nigeria (Cline-Cole, this volume). While in Tiffen's chapter, the 'blame' for these biases and omissions tends to be laid at the feet of the individual economists and anthropologists who are seen to be responsible, the problem may be broader. Below, we examine why such blind spots exist by focusing on the context within which science operates.

Scientific method

Theoretical issues in science are frequently inscribed in the methods which generate supportive data; as such, theory and methodology are hard to distinguish. Nevertheless, a more specific focus on methodological questions is also helpful in comprehending the persistence of orthodoxies about African environmental change. In virtually any discipline particular methods come to acquire credibility and authority, and it can be the inheritance of such methods – as much as of the actual messages they generate – that explains the persistence of some received ideas. By defining what is acceptable as evidence, certain privileged methods also act to exclude other sorts of data. It is in this way that certain questions remain unasked, and certain types of evidence are ignored or dismissed as invalid.

The exclusion of historical data from much ecological science as applied to Africa is a case in point. Orthodox views are often based on speculative projections backwards from the present time, in which present landscapes are presumed to be changed – or degraded – versions of those supposed once to have existed. The 'snapshot' methods on which such views are based came to dominate at a period when time-series data, from air photography and satellite images, for instance, were unavailable to researchers. Now they are increasingly obtainable for periods stretching back several decades, and researchers are becoming increasingly aware of their value. The case studies in this volume illustrate clearly how historical data sets – often combining photographic with official written records, early travellers' accounts, and ethnographic research methods such as oral history – call into question conventional views of environmental change and their methodological underpinnings.

Take the example of botany. The possibility and validity of 'reading' historical process from present form has been strongly developed in 'phyto-sociology'; an approach which uses the present species composition, structure and boundaries of vegetation communities as a basis for deducing the processes which have led to this form. Yet, as Fairhead and Leach (this volume) show with reference to the 'derivation' of savanna landscapes, the assumptions about process in these deductions may prove unfounded. Rather than regarding the mixed forest-savanna species composition of forest-patch boundaries as evidence of 'savannization' and forest degradation, as many botanists in forest Guinea have assumed (e.g. Aubréville, 1949), Fairhead and Leach show it to be the outcome of forest expansion into savanna, resulting from the deliberate management of soil, trees and fire by local farmers.

Time-series aerial photograph comparisons support farmers' own oral-historical explanations that their strategies serve to increase forest cover over the longer term.

Another type of methodological blind spot has been mistakenly to assume conditions at a particular time to be representative of an enduring state of affairs. Brockington and Homewood (this volume), for example, describe how colonial administrators interpreted the low population levels prevailing at the turn of the twentieth century as the norm for East African savannas, when in fact they were the result of the decimation of human and livestock populations by war, famine and disease in the late nineteenth century. Not only did this convey a misleading impression of unpopulated 'wilderness' to early European settlers (cf. Anderson and Grove, 1987), but it also fuelled anxieties about African population growth later in the twentieth century, which proved expedient in politically motivated arguments that the 'carrying capacity' of the land would quickly be exceeded.

An equally common methodological error has been to take short-run observations as evidence of a secular, long-run trend, when they may simply describe one phase in a cycle. The clearing phase in a fallow cycle, for example, need not imply the long-term removal of vegetation. Swift (this volume) describes how received ideas about desertification have rested partly on the observed expansion of desert margins over periods of little more than a handful of years. Analysis of the data for several decades usually reveals that such observations capture only a small wobble in the long-run fluctuation of the desert margin, driven by climatic cycles (cf. Helldén, 1991). Other cyclical factors can also colour outside observers' impressions of range condition at a given point in time. Taking the case of South Africa, Beinart (this volume) shows the significance of the chosen observation period in relation to the long-run dynamics of livestock herd build-up and decline, determined largely by political and economic conditions.

An analogous type of error is the scale problem discussed by Stocking (this volume) in relation to the measurement of soil erosion. Just as short-run observations may give a poor indication of long-run trends over time, so extrapolation from a small-scale erosion plot to an entire catchment area, region or even country gives a wildly exaggerated impression of the real severity of soil erosion. Although the flaws in this approach have been recognized for some time, '[p]roposals for major measurement programmes of soil loss using field plots still appeal' (Stocking, this volume). This is just one of a number of methodological traps commonly encountered in the measurement of soil erosion, which combine to give a highly misleading impression of its rate and severity in Africa. Yet, as they are lent professional authority by the scientific establishment, such assessments have come to represent a chief source of legitimacy for orthodox soil conservation programmes in Africa, such as those carried out in Ethiopia (Hoben, this volume).

We have mentioned here only a few of the examples given by the contributors to this volume of how science and methodology, and the selective use and misuse of the resulting empirical evidence, form central props to received wisdom about environmental change in Africa. This is based not on ignorance, then, but on particular forms of 'knowledge', and what is taking place is the exercise of power through subtle but effective instruments for the formation and accumulation of knowledge, including 'methods of observation, techniques of registration, procedures for investigation and research, [and] apparatuses of control' (Foucault, 1980: 102). The power relations embodied in the production of knowledge are reinforced by the ideas, values and methods of 'normal professionalism' in rural development (Chambers, 1993). It is not merely that 'knowledge itself is power' (Francis Bacon; cited in Davies, 1994: 1), but that 'what constitutes knowledge, what is to be excluded and who is designated as qualified to know involves acts of power' (Foucault, 1971; cited in Scoones and Thompson, 1994: 24). In the following sections we examine how the embedding of such knowledge/power relations in the institutions of science and of development help account for the persistence of received ideas about African environmental change.

Sociology of science in public policy
We noted in the previous section that the origins of many of the orthodoxies we are concerned with here can be traced back to the early colonial period. Both distance and recent research make nineteenth and early twentieth-century imperialism a good choice of historical period for the purposes of illustrating how the broader economic, political and institutional context shape the manner in which science is put to use in public policy.

The political and economic context, for example, helps to explain the differences in the kinds of conservationist policy adopted in the settler colonies of British East and southern Africa, and those adopted in the trading colonies of British West Africa (Anderson and Millington, 1987). An overt, social control agenda lay behind policies in East and southern Africa that placed physical restrictions on African farming activities, supposedly in the interests of conservation, because they directly threatened the interests of European settlers (Anderson, 1984; Beinart, 1984). These restrictions included, for example, the prohibition of farming on slopes claimed to be too steep for cultivation (Showers, 1989), the relegation of African farming to designated reserve areas, and restrictions placed on the crops African farmers were allowed to grow, especially with respect to export crops such as coffee (Tiffen *et al.*, 1994). In all these cases, conservationist arguments were made to justify those policies. By contrast, in the trading colonies of West Africa where African cash-cropping and gathering were important to colonial economic interests, conservationist policies to restrict these activities were much less prominent (Millington, 1987). Thus, in selective ways, received ideas about environmental change were marshalled to justify one or another type of public policy.

Colonial science, as it came to be applied in Africa, was strongly influenced by scientific ideas and debates originating elsewhere. The influence of the North American dustbowl in the 1930s on thinking about soil erosion in Africa has been well-documented (Anderson, 1984; Beinart, 1984; cf. Brockington and Homewood, this volume; Adams, this volume), but there were much earlier precedents. A coherent interpretation of environmental degradation attributed to the economic demands of colonial expansion emerged in the island colonies of St Helena and Mauritius as early as the mid-seventeenth century (Grove, 1995). The notion of a causal link between deforestation and declining rainfall already held wide currency among the medical surgeons, botanists and other scientists employed by trading companies such as the East India Company by the mid-nineteenth century, and through them came to play a decisive role in shaping India's forest conservancy system (Grove, 1989). The Indian system, in turn, provided a model for those in Africa and, much later, North America (Grove, 1995). And early in the twentieth century, the scientific study tours of the Indian-trained Stebbing were particularly influential in emergent African debates about desertification (Swift, this volume).

The extent to which colonial scientists were part of a 'global' network or community was echoed even more strongly within the African continent itself. Scientific tours and visits, as well as regional commissions and conferences, created a context in which individuals were often able to influence entire regions and generations. For example, Pole-Evans influenced a whole generation of rangeland scientists in southern Africa and beyond (Scoones, this volume); and the interactions of the francophone botanists Chevalier and Aubréville with Nigerian and Ghanaian botanists and foresters, particularly Thompson and Unwin, ensured that their analyses of vegetation were carried over the anglophone-francophone divide in West Africa (Fairhead and Leach, this volume). The written texts and maps of such influential figures frequently became key reference works for subsequent generations of scientists in comprehending African environments. It is little surprise, then, that a remarkable consistency in ecological analysis often emerged and persisted across whole ecological zones.

While the terms of scientific debate during the colonial period appear to exhibit remarkable consistency, it would be misleading to portray scientific opinion at any given moment as homogeneous. For example Beinart (this volume) describes the strong debates around assessments of grassland condition in South Africa between the 1930s and the 1950s. Unlike his contemporaries, the botanist C.E. Tidmarsh was rather cautious about the orthodox view of continuous grassland degradation, and attributed grassland composition more to climate and available moisture supply than to the nature of grazing treatment. Tidmarsh's arguments predated the recent literature on non-equilibrium range ecology (Behnke *et al.*, 1993) by some forty years. However, although Tidmarsh served on the Desert Encroachment Committee, his ideas had little lasting influence on public policy. Similarly, in the context of southern US agriculture between the seventeenth and nineteenth centuries, ideas about environmental change, agricultural innovation and the role of indigenous farmers appear to have fluctuated on a cycle of several decades (Earle, 1988). Although the mainstream view of agriculture during the period is characterized by the notion of the southern farmer as 'soil miner', Earle seeks to explode this myth by highlighting those periods in which 'folk' innovations in agriculture were viewed positively by outside observers, in a manner reminiscent of the contemporary 'farmer first' paradigm in agricultural development (Chambers *et al.*, 1989).

The financial and bureaucratic structures of scientific establishments and public administrations strongly influence these processes, shaping debate – and freedom to pursue it – at any given moment. In many cases, those whose ideas posed a significant challenge to the prevailing orthodoxy found their views either suppressed or unable to influence higher levels in the institutional hierarchy. For instance, Fairhead and Leach (this volume) show how such constraints have operated through colonial, post-independence and recent times to stifle challenges to the dominant view of vegetation dynamics in Guinea's forest-savanna zone.

Even changing and contested views of environmental change in Africa may have posed little real threat to the continuity of policy and practice. Adams (this volume) suggests that changing sets of ideas originating at the centre may, in the course of their transmission to the more distant outposts of public administration, 'sediment down' more slowly and adapt themselves to preceding thinking. There can, therefore, be a remarkable degree of continuity in what happens on the ground, as he shows with reference to the rationale for external intervention in irrigation in Marakwet district of Kenya from the 1930s onwards. Adams describes how colonial administrators could at the same time be impressed by indigenous irrigation furrows, yet regard them as a cause of soil erosion and local irrigators to be incapable of managing them properly. Whatever view prevailed of indigenous irrigation practice was consistently manipulated in order to justify European intervention and imposed change.

Many of the colonial scientists to whom we have referred as having a disproportionate influence over the early origins of received wisdom about African environmental change were not only scientists. In most cases, they were employed as public servants, and played decisive roles in colonial policy formulation and administration. The botanist Pole-Evans, for example, is also credited with having shaped the draconian Natural Resources Act (1942) in Rhodesia and the Swynnerton Plan

(1954) for agricultural intensification in Kenya (Scoones, this volume). In French West Africa the botanist Aubréville eventually rose to become Inspecteur Général des Eaux et Forêts des Colonies in French West Africa (Fairhead and Leach, this volume). And in Nigeria, the forest conservation enthusiast Moloney came to be Governor of Lagos Colony (Grove, 1994), while Lugard (Cline-Cole, this volume) established a regional forest service in Northern Nigeria.

Early in the colonial period, these individuals had little evidence to support their hypotheses about African environments. Nevertheless, such hypotheses became institutionalized in the colonial agricultural, forestry, livestock and wildlife departments, forming the rationale for intervention. Thus even if the scientific analysis to provide empirical support to early contentions about the relationships between rainfall and deforestation, or stocking density and range condition, had not yet been carried out, the agenda for such analysis was already set through the establishment of these institutions. And in turn, the persistence of these institutional structures provided a context in which their analysis could remain dominant, and be further elaborated.

At least in some parts of Africa, the colonial legacy in environmental institutions was directly inherited by post-independence governments, helping to account for the persistence of received wisdoms. Furthermore, by comparison with the colonial period, some notable similarities can be observed in the relationship of contemporary expatriate scientists and academic advisers to the process of public policy formation in Africa. One is the exchange of ideas within a network or community of like-minded individuals; a second is the tendency for scientists to be – more or less directly – 'in the pay' of policy institutions.

Sociology and practical effects of development
The foregoing arguments begin to suggest how received wisdoms about environmental change are institutionalized. We now turn more directly to institutional issues, and examine how structural factors in the contemporary development process itself help explain the persistence and power of particular views of African environments. Of key importance here are the inextricable links between development institutions, their analyses and the effects 'on the ground' of the policies they promote; effects which have often proved detrimental from the perspectives of local land users.

It is possible to show that the interests of various actors in development – government agents, officials of donor agencies, the staff of Northern and Southern non-governmental organizations, and independent 'experts' – are served by the perpetuation of orthodox views, particularly those regarding the destructive role of local inhabitants. Pejorative attitudes and repressive policies towards pastoralists, for example, have been well-served by the view that they cause desertification, or bring about a tragedy of the commons (Swift, this volume; Scoones, this volume). And in East Africa, it suited the land-expansionist concerns of white settlers to attribute soil erosion to the 'primitive' practices of indigenous farmers (Anderson 1984; cf. Adams, this volume).

Throughout the colonial period, received wisdom about environmental change served to justify the formation and funding of national-level executive agencies with responsibility for environmental management. This has continued to the present day. Government departments with responsibility for forest and wildlife protection and management, in particular, are often heavily reliant on revenues received from fines and the sale of permits. The underlying premise on which the continued flow of such revenues rests is that stewardship over natural resources is properly the responsibility of the state. It depends on and serves to perpetuate the conventional view that local inhabitants are incapable of acting as resource custodians. In order to justify the existence and expansion of natural resource departments as an arm of state administration, therefore, there are strong vested interests on the part of government agents in maintaining received wisdom about the instrumental role of local inhabitants in bringing about environmental degradation (Tiffen, this volume; Fairhead and Leach, this volume).

Common to almost all the contributions in this volume is the view that received wisdom about African environmental change has had the instrumental effect of promoting external intervention in the control and use of natural resources. In a bold statement of this position, Roe states that

> . . . crisis narratives are the primary means whereby development experts and the institutions for which they work claim rights to stewardship over land and resources they do not own. By generating and appealing to crisis narratives, technical experts and managers assert rights as 'stakeholders' in the land and resources they say are under crisis (Roe, 1995: 1066).

He goes on to argue that shifts in narratives can reinforce, rather than undermine this process, since

> . . . the more crisis narratives generated by an expert elite, the more the elite appears to have established a claim to the resources it says are under crisis . . . whether right or wrong, the claims, counter-claims and changing claims of experts serve principally to reinforce and widen the belief that what they, the experts, have to say really matters and matters solely by virtue of their expertise (*ibid.*)

More broadly, others suggest that particular kinds of development discourse serve to justify the expansion of bureaucratic power in rural areas. For example, Ferguson (1990) documents how the bureaucratic logic of the Thaba-Tseka Development Project in Lesotho served to depoliticize poverty and powerlessness so that they could be portrayed as a set of 'technical problems' awaiting solution by 'development' experts. Drawing on the work of Foucault, but sharing much with the analysis of Wood (1985) and others, Ferguson reveals the planning apparatus and the conceptual system on which it rests as mere cogs in the development 'machine', linking up with social institutions in such a way as to shape outcomes without actually determining them (cf. Long and Long, 1992). The principal outcome, Ferguson argues, is to promote, almost unnoticed, the pre-eminently political operation of strengthening the state presence in rural Lesotho.

The process Ferguson describes is not uncommon in planned interventions in natural resource management. His analysis gives an insight into the seemingly inexorable way in which it is convenient for government and donor agencies to promote particular, off-the-shelf intervention 'packages' which frame problems and solutions in technical terms, obscuring alternative analyses such as those which might lie in the realm of political economy. Amanor (1994a), for example, describes how in the forestry 'sector' such packages may take the form of replicable 'green technologies' such as alley cropping, or of organizational packages such as those promoted under the Tropical Forestry Action Plan (FAO, 1985). Several contributions in this volume describe the application of standardized intervention packages for soil conservation (Stocking; Hoben; Adams), livestock management (Scoones; Brockington and Homewood) and forestry (Fairhead and Leach; Cline-Cole).

As illustrated by many of the cases here, such external claims over resource management and control can have deleterious consequences for local livelihoods. They can marginalize or alienate people from natural resources over which they previously enjoyed access and control, perhaps directly undermining their ability to secure food or income. This has sometimes been the case, for instance, with policies to exclude people from externally managed forest or wildlife reserves, or to confine pastoralists to fenced paddocks. Where inhabitants must, out of necessity, continue to use resources claimed by external agencies, they often find themselves subject to taxes or fines which render them more resource-poor. Even when inhabitants retain rights to use resources, the imposition of external regimes or 'packages' for their management can impose unwelcome demands on their labour and resources. Hoben (this volume), for example, describes how Ethiopian highland farmers resented the costs in labour, time in meetings and so on expected of their 'participation' in soil conservation projects which they knew to be inappropriate.

In some cases, the assertion of professionalized claims over land and resources has also had adverse ecological consequences. For example, Fairhead and Leach describe how external prohibitions on the setting of bush fire undermined inhabitants' early-burning strategies, risking greater fire damage by late dry season fires. By contrast with negative effects on local populations, which may be of little concern to – or even in the interests of – certain state agencies, such counterproductive environmental effects might be thought to throw policy approaches and their supporting analysis into question. Yet an effect of labelling in the framing of technical problems, as Wood (1985) notes, is to contribute to a self-fulfilling cycle of policy failure which deters such questioning. As Scoones (this volume) shows with reference to rangeland management in Zimbabwe, if a given policy or intervention package fails or is resisted by the 'target group', it is implicitly assumed to be because the 'target group' is recalcitrant or ignorant, rather than because the problem was misconceived in the first place. As a result, efforts are redoubled on the part of the implementing agency to bring about the same desired outcome, but with the use of greater force, which merely serves to worsen the initial 'policy' failure.

In this light, it is not surprising that policies have often been resisted by local people. Such resistance needs to be interpreted both as a response to social repression, and to inhabitants' understanding that policies were often inappropriate to local ecological conditions. Resistance sometimes takes 'everyday', covert forms (Scott, 1985), including coping and adaptive strategies to carry on with customary practices regardless of the consequences; and sometimes takes more overt forms of political expression.

Nevertheless, it would clearly be wrong to portray local inhabitants only as victims of repressive colonial conservation policies. Evidence from West Africa, for example, suggests that local elites, at least, were quite successful in subverting external forest policy agendas, and in turning them to their own advantage (Grove, 1994). Furthermore, the contributors to this book are careful not to overplay the significance of planned development in analysing the practical effects on the ground and in people's lives of the received wisdom under scrutiny in this book. Changes in people–environment relations certainly do not come about only through development policy, while planned interventions are simply part of a chain of events within a broader framework of activities of the state and various interest groups (Long and Van der Ploeg, 1989), as Tiffen's contribution to this volume shows in tracing the diverse sources of capital in Kenyan rural development. Heeding Clay and Schaffer's (1984) warning on the artificial separation of policy and implementation in development practice, Long and Van der Ploeg remind us that interventions merely 'come to form part of the resources and constraints of the social strategies' devised by individuals and 'target groups' affected by development (1989: 228).

If orthodox thinking about natural resource stewardship provides the *raison d'être* for certain state institutions, it is relevant to ask whether it might change along with a change in political context. How much room for manoeuvre is there to shift the environmental agenda? Periods of transition from one political regime to another may provide an opportunity for discourses of resistance to be voiced more strongly. During pre-election periods in West Africa, for example, opposition parties commonly attempt to gain support by expressing discontent with repressive natural-resource 'policing' activities on the part of the state, as illustrated for Nigeria by Cline-Cole (this volume) and for Guinea by Fairhead and Leach (this volume). The question remains, however, to what extent this actually affects the substance of environmental debate. In the case of South Africa, for example, Beinart (this volume) argues that in reversing the direction of blame from African farmers to white farmers, anti-apartheid literature merely turns conventional arguments about environmental degradation on their heads. It thus reinforces dominant readings of environmental history, rather than begins to challenge them.

In contemporary development practice, other actors and institutions have joined the state and national elites in perpetuating received thinking about environment–society relationships. One such group are donor agencies. Hoben (1995: 1,009) suggests that, 'regardless of its merit, an environmental paradigm is transferred to aid recipient countries through training, institution building and investment. These activities attract and create elite interest groups which, in turn, become its constituency, making it politically difficult to discard.' He adds to this a number of other conditions relating to donor agencies which contribute to the entrenchment of a given environmental narrative. They include the dependence of weak African government departments on official development assistance; and the political and moral pressures on donors to be seen to respond to their domestic constituencies and to act quickly (Hoben, this volume). These conditions create a policy-making environment within which neo-Malthusian, 'crisis' narratives, in particular, can flourish.

The reliance of African governments on foreign assistance for environment-related 'development' activities is by no means new. In the colonial period, for example, following the attention generated by Africa-wide environment conferences, funding envelopes often became available for which colonial administrations could apply. But the late 1980s and early 1990s saw the onset of 'green conditionality' (Davies, 1992) as donor agencies began to use environmental goals as a form of leverage over national governments in the process of 'policy dialogue' (Leach and Mearns, 1991; Davies and Leach, 1991).

Since the 1992 Earth Summit new international financial mechanisms have emerged to address environmental problems conceived at a global level. The Global Environmental Facility, for example, jointly implemented by the World Bank, UNDP and UNEP, is concerned with the loss of biodiversity, climate change, the pollution of international waters and ozone layer depletion. In this context, conformity to globalized, commonly held conceptions of environmental problems has become important for local environmental activities to attract funds.

At the same time, non-governmental organizations (NGOs) have become increasingly important actors in the international development community, and have also contributed to the 'greening of aid'. Although in the 1990s multilateral and bilateral donor agencies are channelling an increasing volume of official development assistance through NGOs, the activities of many Northern-based charitable organizations have long relied on fund-raising from the general public. Their campaigns must therefore appeal to and elicit a response from a wide audience. Ironically, this often serves to reinforce stereotyped images which, in their development education mode, the very same organizations may wish to challenge. The Northern popular perception of famine as evidence of a Malthusian crisis in Africa with an environmental dimension, for example, played an important role in sustaining aid flows to Ethiopia in the mid-1980s (Hoben, this volume), but arguably did little to further understanding among the general public in the North of social and political realities in Africa.

These contradictory relationships are replicated in the alliances between Southern NGOs and the Northern NGOs on which they are financially dependent (Hudock, 1995). In order to attract funding, Southern NGOs must respond to and comply with an environmental agenda set by their Northern partners and, in so doing, can reproduce a view of local land users which denies the perceptions and interests of their own local constituencies. They tend to internalize the specific discourse used by Northern NGOs to justify funding environment-related activities, and in many cases even owe their very existence to it. For example, Fairhead and Leach (this volume) suggest that the emergence of the urban-based 'Friends of Nature' societies in Guinea, or at least the discourse they adopt, can be attributed to the availability of external funding tied to a specific environmental agenda. Received wisdom about environmental change and human agency and the *raison d'être* of many Southern NGOs can therefore be mutually reinforcing.

Another relative newcomer to the contemporary international development community is the independent consultant 'expert', whose influence over the shape of development policies, projects and programmes in Africa is now indisputable. This actor plays a unique role in the reproduction of environmental orthodoxies which has to do, we argue, with the nature of accountability. As Tiffen remarks (this volume), academics and civil servants who advise donor agencies and African governments on agricultural and other forms of development policy are accountable not to those who are the intended beneficiaries – Africa's farmers – but to public sector agencies as their paymasters. This kind of 'backwards accountability' had its origins in colonial service, since colonial scientist-administrators were similarly accountable to the higher echelons of the civil service to which they aspired. The independent consultant of today, however, is faced with rather different incentives and pressures.

The reduction of international aid budgets in the 1990s has led donor agencies increasingly to use independent consultants rather than those from public bodies and private firms whose fee rates are inflated by institutional overheads. Independent consultants, it could be argued, are accountable only to their own curricula vitae. The terms of reference for short-term contracts tend to be pre-set so that, for example, the consultant is required simply to describe the social causes and consequences of a particular environmental problem without ever questioning its existence. Under strong time pressure, an analysis is produced which tacitly confirms and further reinforces the conventional wisdom embodied in the original terms of reference. Even if consultants are well aware that the underlying premises of their work may be called into question, the incentive structure is unlikely to lead them to operate differently. Indeed, the market for consultancy services is so segmented that even the risk of a damaged reputation in one field or geographical area, or with one multilateral or bilateral agency, need not unduly hinder an individual's chances of winning new contracts elsewhere in future. None of these remarks are intended to suggest that all independent consultants behave irresponsibly, or to deny instances where committed consultants have strongly criticized conventional views. There seems little doubt, however, that this new group of actors can play a significant role in fixing environmental narratives.

Popular culture and the social construction of environmental meaning
We do not suggest that received wisdom can be explained simply in terms of the interests of these or any other individual set of actors. It is not the case that any single group deliberately conspires to engender environmental or other messages that best suit their interests. It is more that through their continuing interactions with others, individual actors can unwittingly participate in the social construction of particular forms of environmental meaning. Here we explore further how such meanings can resonate with existing symbols and meanings in popular culture such as to produce and reinforce received wisdom.

The 'environment' constitutes not one single issue, but many diverse ones which 'do not ordinarily articulate themselves' (Hansen, 1991: 449). Claims therefore have to be made by various actors about what constitute environmental issues for social concern. This process often begins in the scientific forum, but its subsequent inflection takes place through complex interaction with other arenas for public debate, including courts of law, formal politics, and the mass media. The mass media do not simply transmit messages to their audiences about 'the real world'. Rather, they participate in the social construction of environmental problems by articulating culturally specific and encoded 'messages', which are subsequently decoded and given meaning by their audiences within existing frames of reference (Burgess, 1990). Environmental understanding is continually being transformed by the interactions of all the participants in this process.

Research on the role of the media in shaping environmental understanding has addressed both the volume of environmental coverage in the mass media (e.g. Lacey and Longman, 1993), and the character of that coverage (Lowe and Morrison, 1984). While noting 'the journalistic preference for the negative and the dramatic' in news in general, Lowe and Morrison point out that 'it is especially difficult to feature the positive within environmental reportage' (Lowe and Morrison, 1984: 78). There appears to be an inbuilt tendency for the media to generate 'crisis' narratives with respect to environmental issues.

Owing to the globalization of information flows in recent years, the range of actors who now play a role in producing and consuming ideas about environment and development is historically unprecedented. The rapid technological development of communications media, including satellite television and the internet as well as more conventional forms of mass media such as terrestrial television, radio and newspapers, means that the global circulation of information has never been greater or accessible to a wider audience (Davies, 1994). This has led to new and complex forms of claims-making for environmental issues, in which popular culture and the mass media play especially important roles. As Burgess argues,

> The power to define the meanings of landscapes and places, plants and animals, renewable and non-renewable resources is being contested in new and fascinating forms of cultural politics conducted primarily through the mass media: take, for example, the alliance between actors, musicians, Brazilian Indians, pop music promoters, conservation organizations, the media industry and the mainly young consumers who buy records to support the campaign against the destruction of the Amazonian rainforest (Burgess, 1990: 141).

Paradoxically, this more rapid circulation of information may actually increase the tendency towards simplification and convergence in the substance of popular discourse about environment and development, as a way of dealing with information overload. Public debate about environment and development issues necessarily involves other actors in the development process too, and contributes to the construction and simplification of environmental messages. Political pressure is brought to bear on multilateral and bilateral donor agencies through non-governmental organizations and other domestic constituencies, but the terms of debate are increasingly global and symbolic. The same images, the same often-repeated statistics that define environmental change, are frequently internalized and reproduced by scientists and administrators working at the local level.

While the inhabitants of local African environments may themselves participate in the production of ideas about environmental change, they do so with less power to define the terms of debate. As token participants in global and national fora, they may have little chance to express alternatives to the dominant viewpoint. But equally, it is not uncommon for rural inhabitants in their interactions with development fieldworkers to confirm outsiders' preconceived ideas, given the power relations which operate at such 'interfaces' (Long, 1989; Long and Long, 1992). Such confirmation may arise out of fear, suspicion, or a desire to remain on good terms by accepting what is being offered, as well as the relations of authority and the memory of past experience which structure these interactions. All too often, the power relations which shape such encounters remain invisible to the outsiders involved, leading them blindly to accept local accounts as indisputable. By attempting to increase dialogue with local people, some efforts to enhance 'participation' in development may, ironically, contribute to the very conditions which permit such misunderstandings to occur.

More significantly, farmers and herders may also selectively adopt outsiders' environmental idioms and turn them to their own advantage in struggles over identity and resource control. Such expression of 'environmental' discourse may bear little relation to local people's own, practical ecological knowledge. For example in Guinea, Fairhead and Leach (this volume) describe how externally derived images of forest loss are invoked by Kissi and Kuranko villagers in discourse about ethnicity, to identify themselves respectively as 'forest people' or 'savanna people' in ways which – in colonial and now modern Guinea – have political significance. Thus 'there is no way of keeping the conceptual apparatus of the observer . . . from appropriation by lay actors' (Giddens, 1987: 19; cited in Long and Long, 1992: 39), although such appropriation may respond more to other pressing concerns in popular culture than to 'environmental' concerns *per se.*

To summarize, we have argued that the reasons for the origins and persistence of received wisdom about environmental change in Africa lie in the substance of science, on the one hand, and in its social and historical context, on the other, including the effects that it has through development in practice. To the extent that science is often the 'primary definer' (Hansen, 1991) of what constitute environmental problems, it is relevant to ask how that science is carried out. Brockington and Homewood (this volume) suggest that 'good' natural science can be used to judge between competing social science explanations for dryland degradation. Swift (this volume), on the other hand, asks why it is that in the debate about desertification the results of 'poor' science tend to be picked up and used more often than those of 'good' science. Sometimes the orthodox view appears to persist because it is politically expedient to suppress or ignore evidence to the contrary. From this viewpoint, science and the broader political economy are regarded as distinct from one another. Alternatively, 'evidence' itself is seen to be produced through a discursive process in which scientists are just one set of actors among several. This view, held by sociologists of science, emphasizes the simultaneous construction of knowledge and social commitments.

Both forms of explanation are represented in this volume. On the one hand there are structuralist explanations, within a rational choice or 'political economy' framework. The perpetuation of received thinking about desertification, for example, seems quite clearly to serve the political or economic interests of relatively powerful groups in donor agencies and recipient governments, with the tacit support of sections of the scientific community (Swift, this volume). On the other hand there are 'actor-network' explanations (cf. Long and Long, 1992). In these cases, the 'stickiness' of received wisdom is explained by the convergence of ideas and social commitments on the part of various actors, including local inhabitants, at particular historical moments (Fairhead and Leach, this volume). An additional layer to such analysis suggests that there are hierarchical relations of power between various participating actors, which leads such convergences of commitments to coalesce in certain, dominant directions (cf. Foucault, 1980). It is this, we argue, that accounts for the remarkable historical continuity in received wisdom about environmental change in Africa.

Ways forward in research

The cases here show that the policies founded on environmental orthodoxies have often proved not merely harmful to African farmers and herders, but ineffective in ecological terms as well. Given the power relations through which, as we have argued, orthodoxies are produced and sustained, there is clearly no simple remedy for this state of affairs. Nor is it likely that 'more and better research' could improve the outcomes of policy for Africa's farmers and herders without more fundamental changes in the relationship between research and development policy-making, and between the diverse institutions which influence policy processes. The issues involved here extend far beyond the scope of this book. Nevertheless, the cases assembled here do point towards ways in which applied research could be improved. In this last section we reflect on some of these research implications, moving from the level of place-specific analyses of society–environment relations, to broader issues concerning the role of research in the policy process.

Numerous contributors show how alternative analyses of environmental change and people's roles in it imply very different kinds of policy from those suggested by received wisdom. For example, Scoones' and Swift's contributions join a major rethinking of pastoral development policy which has followed new understandings of rangeland ecology (Behnke *et al.* 1993; Scoones, 1995). If pastoralists' herd and land-management strategies are not, in fact, precipitating overgrazing and linear degradation, but instead are making the most of productive opportunities in highly variable and patchy dryland environments, then there are good grounds for adopting policies that support those strategies, for instance through flexible land-tenure arrangements which allow herders to maintain mobility.

In Guinea's forest-savanna transition zone, Fairhead and Leach's reconsideration of vegetation history suggests specific technical avenues for interventions in the field of natural resource management. Many aspects of local farming and everyday resource use have historically and presently served to increase forest cover, whether by manipulating particular tree species, vegetation communities or soils to 'deflect' savanna to forest successions. Such 'integrated forest management', appropriate to local ecology and labour conditions, could provide an effective basis for external support; while the analysis suggests that policy makers should be encouraging, rather than repressing, key aspects of inhabitants' upland farming and fire management (cf. Leach and Fairhead, 1994).

In highland Ethiopia, Hoben's alternative analysis of the causes of soil erosion suggests very different policy solutions from those contained within the orthodox, neo-Malthusian narrative. If degradation is the product largely of political and economic conditions unfavourable to soil conservation, then appropriate solutions are more likely to lie in the realm of institutional and tenurial reform, rather than in attempting to educate and force farmers to adopt erosion control techniques of which they may already be aware.

In their implications for policy, these countervailing analyses and others explored in this book share certain common elements. One is that they emphasize working with and building on the ecological knowledge and skills of Africa's farmers and herders; the very skills often rendered invisible by neo-Malthusian degradation narratives. In particular, they suggest that local inhabitants may have long been practising 'opportunistic' resource management attuned to non-equilibrium ecological conditions. A second, related emphasis is on creating the enabling conditions under which local resource-management strategies can be pursued effectively. Several contributors draw attention to people's own 'investments' in enriching their land and intensifying its use, provided prevailing institutional and economic conditions (such as tenure regimes, marketing arrangements and other aspects of infrastructure) make this possible and worthwhile (e.g. Cline-Cole; Fairhead and Leach; Tiffen).

Many of the contributors are careful to note the place-specificity of their alternative analyses, and would therefore make no claims that their policy implications necessarily extend throughout the ecological zones or natural-resource management 'sectors' in question. However, it would seem unwise to treat these cases as exceptions until it has been proved that their analyses do not apply elsewhere. In vast areas of Africa's drylands, highlands and forest margins, similar investigations have yet to be carried out. To date, received wisdom about environmental change has been so taken for granted that deeper interrogation has seemed unnecessary. The cases thus imply an urgent need for further research elsewhere to explore findings such as these in a comparative way; research which makes use of historical and time-series data sets; which pays serious attention to inhabitants' own experiences and opinions, and which opens its questions to issues in 'new' ecology.

There are research implications here too for ecological science, in which theory as well as empirical tenets have begun to be re-cast in recent years with the adoption of historical approaches. The insights from these cases invite more widespread studies of ecological history that test alternative theoretical propositions even where issues may seem cut and dried. This, in turn, demands openness and willingness on the part of scientists not only to explore radical counter-hypotheses about environmental change – including those which stand conventional wisdom on its head – but also to the possibility that indigenous ecological knowledge and opinion might provide such counter-hypotheses. Such a research agenda clearly presents methodological challenges, especially to use the historical data sets now available in innovative ways. Several cases in this volume illustrate the promising scope in a judicious combination of air photographs and satellite imagery with archives and detailed, ethnographic research.

Yet, as the analysis in this introduction indicates, 'better scientific research' is unlikely to have practical impact on orthodox thinking and practice without more fundamental changes in the institutional structures through which environmental problem-claims are made, solutions elaborated, and translated into policy and practice. The ineffectiveness of shifts in one set of 'props' to received wisdom, without changes in the others, is illustrated by several of the cases in this book. As Scoones describes, a strong body of scientific evidence and the convictions of a large group of committed scientists and development practitioners now support alternative approaches to rangeland management in Zimbabwe, yet it remains highly uncertain whether these approaches will prove politically acceptable to the state and development agencies, or implementable given institutional commitments to established policies of external regulation. The situation described by Adams, in contrast, illustrates how received wisdom within policy institutions, as well as among scientists, can shift quite dramatically, but does so largely in line with shifts in dominant political interests (in this case, those of white settlers, famine relief, and so on). Under both sets of circumstances, shifts in thinking proved insufficient to give voice to inhabitants' own opinions and experiences in a way that could effectively be translated into changed policy and practice. Furthermore, Adams' analysis suggests that an approach based on 'more research' can only go so far, and may ultimately be self-defeating, since there will always be more rounds of orthodoxies and counter-orthodoxies.

It is not merely the case that received wisdom and challenges to it draw different conclusions about environment and people's interaction with it; they also uphold different social and political commitments and claims. Even in the cases where received wisdom can be shown to be demonstrably false, there will not necessarily be a single alternative analysis which can be shown to be 'truer' to all parties involved, nor policy solutions derived from it which all would find acceptable. As Wynne (1992a) suggests, reflecting a view now widely held by philosophers of science, all knowledge is conditional, in the sense that it reflects the institutional context in which it is produced. Specific readings of environmental change should therefore be treated less in terms of their claims to 'truth', and more in terms of the implicit social commitments that underlie them and on the validity of which they depend. Several such readings or 'plural rationalities' may thus co-exist, and it is the job of good public policy research to make them explicit enough to debate (Thompson, 1993). This task, however, requires a radical shift in the relationship between 'research' and 'policy-making' as conventionally conceived.

Many authors have made the case for such a re-conceptualization of the policy process in complex situations characterized by insurmountable uncertainty (Thompson *et al.*, 1986) or indeterminacy (Wynne, 1992a). The analyses in this book of changing environment–society relations in Africa reveal similar degrees of uncertainty associated with knowing the character, direction and strength of causal linkages, and with designing policies that achieve their stated objectives once put into practice (cf. Mearns, 1991). Under such conditions of uncertainty, conventional, managerialist policy blueprints are of questionable validity. This is especially true in cases of epistemological uncertainty, or 'ignorance-of-ignorance' (Funtowicz and Ravetz, 1992: 259), in which policy makers are unaware that there are things of which they are unaware, including the unintended consequences of a given policy instrument.

From this perspective, the task of linking research into policy-making shifts to one of broadening the range of problem-definition claims, and negotiating outcomes among an extended peer group of actors. These may include scientists, policy makers in governments and donor agencies, local administrators, others such as non-governmental organizations with a stake in environmental protection, the mass media and, of course, local inhabitants themselves. This is tantamount to a 'democratization of expertise' (Funtowicz and Ravetz, 1992).

This reconceptualization of the research-policy process is very much a frontier in the contemporary practice of development. The line taken in work on science and public policy is kindred with the agendas of political ecologists, who express the need 'to help uncover the discourses of resistance [to received wisdom], put them into wider circulation, create networks of ideas' (Peet and Watts, 1993: 247). There are not yet any proven models for practical success in such a reconceptualization, although a powerful case in its favour has been made in a number of contexts. Among these, for example, are the reinterpretation of the causes of environmental degradation in the Himalayas by making plural rationalities explicit (Thompson *et al.*, 1986), and the lessons learned from community-based approaches to wildlife management in Africa, emphasizing an action-research approach to programme design involving local communities, national research institutions, local and national government, and non-governmental organizations (IIED, 1994).

Three types of criticism of a more pluralistic, or democratic, approach to the research-policy process may be anticipated: that it is methodologically weak or unproven; that it is populist or politically naive; and that it generates findings that are too complicated to be of practical use to policy makers (cf. Wynne, 1992a).

While allowing for serious consideration of the knowledge, experience and opinions of 'lay' actors in the development process, those advocating the 'democratization of science' do not imply that 'anything goes', or that the value of conventional scientific research is thus contested. The findings from such research are merely placed on a more provisional footing (Funtowicz and Ravetz, 1992; Pretty, 1994).

On the charge of populism or political naivety, it may be argued that building consensus among actors whose world views and political interests are incommensurable is impossible; that exchanges between policy actors with radically different endowments of power and resources could never be politically neutral. The 'democratization of the policy process' calls to mind the metaphor of an African palaver tree, in whose shade scientists, policy makers, donors and farmers or herders would argue their respective cases, and attempt to come to some agreement. But as participants in African village palavers are well aware, such fora are neither open nor neutral; cases are expected to be won by the most politically influential party; they are expected to invoke particular interpretations of history in their favour, and 'consensus', if reached, may be in little more than outward appearance (cf. Murphy, 1990). Furthermore the real history of political interaction between different actors would certainly condition any attempt to foster more 'democratic' encounters. And in many cases, fundamental differences in the way the 'environment' is valued and plural rationalities of environmental change may be so deeply rooted in their social and cultural contexts that participants in such encounters would be likely to be defeated in their attempts to comprehend and respect each others' perspectives. It is difficult, for instance, to imagine how consensus might be reached where Northern conservationists aspire to empty high forest, and farmers to convert the same land to the bush fallow they value as being more productive. All groups of actors, whether scientists, policy makers or local inhabitants, may be expected to resist such a pluralistic – even relativistic – view of knowledge. Yet a 'democratic' approach to the research-policy process aims precisely to reveal the hidden social and cultural assumptions underlying apparently incommensurable world views. Rendering such conflicts explicit may enable them to be addressed more openly, rather than remain concealed in hegemonic environmental readings and policy.

The third charge against a more 'democratic' approach to the research-policy process is that such an exposure of plural viewpoints would serve to replace a simple received wisdom with excessive complexity. In this context Roe (1991, 1995) argues, pragmatically, that simplified but compelling narratives are in fact necessary to the policy process as currently existing. The challenge is thus to create equally compelling counter-narratives which better fit the claims of a different set of stakeholders; preferably, counter-narratives with equally attractive slogans and labels. In this context, it is recommended that researchers consider ways of working more closely with the mass media in trying to counter received wisdom about environmental change in Africa. A further justification for generating counter-narratives is that excessive plurality or complexity could leave the door open either for no policy at all; or for politically motivated, draconian measures, in the absence of other clear guidelines. While devising counter-narratives may seem expedient, given the current nature of much development policy-making in Africa, it tends to perpetuate the binary-oppositional type of policy debate which has so frustrated attempts to move beyond received wisdom. Most of the contributions to this volume imply, rather, the value of a committed attempt to engage with plural rationalities concerning environment-society interactions, in all their diversity.

These arguments suggest the strong need for changes in the policy process and its institutions, as well as in research. If scientists, policy makers and local inhabitants are genuinely to comprehend each other's perspectives and exchange viewpoints, then innovative institutional arrangements will be needed. If research which reveals a plurality of perspectives is to be useful in the policy process, that policy process and its institutions have simultaneously to change. As a first step in this process, this book demonstrates the importance of revealing the alternative perspectives which challenge received wisdom, and putting them into circulation.

2

Range Management Science & Policy

IAN SCOONES

Politics, Polemics & Pasture in Southern Africa

Introduction[1]

Public policy on rangeland and livestock development in southern Africa remains informed by a particular set of understandings about range ecology and livestock management. This has a long historical lineage, stretching back to the early part of this century. While there may be some emerging cracks in this edifice, the dominant view is based on the science of ecological calamity and stresses the damaging potentials of livestock grazing, the threats of degradation and desertification and the need for control of livestock numbers and grazing movement. This chapter examines the historical origins of this perspective on resource management, exploring the social, political and economic contexts of views on range and livestock management in southern Africa in an attempt to assess why this analysis has remained so resilient to challenge, despite scanty evidence to support it.

The chapter focuses on changes in scientific and policy thinking over the last 60 years. Much of this period has been characterized by both local and global debates about environmental collapse, generating in their wake politically charged and emotive language about the consequences. Policies and plans emerging from these debates often appear to become immune to critical analysis and debate, taking on the appearance of 'truths', resistant to empirical challenge.

The chapter starts by exploring the evolution of thinking in range management science in southern Africa by examining the influence of two eminent scientists on the emergence of a mainstream view, or conventional wisdom, during the 1930s. In the next section, the way this early thinking has shaped public policy in Zimbabwe is explored.

[1] Various colonial and more recent terms are used interchangeably in the text depending on the period being discussed. For instance, in Zimbabwe (formerly Southern Rhodesia), the communal areas were previously known as Tribal Trust Lands or native reserves. Alternatives for place names are placed in brackets in the text, except in the case of countries.

The evidence for the dominant view about environmental change is discussed next, before an assessment of the sources and contexts for challenges to the mainstream view is given. Drawing on this case material, the final section of the chapter looks at how received wisdoms in policy are created, maintained and, in some instances, undermined.

Science and scientists: the emergence of the mainstream

This section focuses on the role of two influential scientists in the establishment of mainstream range science in southern Africa. While they represent only two of many other scientists and administrators working in the region, the case studies serve to illustrate how individuals, linked to institutions and organizations, through professional and other networks, can have enormous impact on the development of ideas. In order to understand how mainstream ideas and practice emerged, I will therefore concentrate in this section on the period prior to the 1940s, a time when many of the key institutions were established, which subsequently helped shape the future of range management in southern Africa. Later sections explore, for the Zimbabwe case, how these ideas were translated into policy and implementation.

Case study 1: Illtyd Pole-Evans
Pole-Evans was born near Cardiff in Wales in 1879 (Gunn, 1968). Following the completion of degrees at Cardiff and Cambridge, he was appointed to the position of Mycologist and Plant Pathologist in the new Transvaal Department of Agriculture in Pretoria in 1905. Starting from scratch, he built up a programme of research. Very much in the tradition of early naturalists who were employed in the colonial service, he was a keen botanist and was involved in extensive collection and documentation of South Africa's aloes and cycads as well as various collection expeditions and surveys of the southern African flora.[2] Through his initiatives the Prime Minister, General Botha, agreed to a botanical survey of the country and the first memoir appeared in 1919.

Pole-Evans was a passionate advocate of vegetation conservation. In his Presidential address to the South African Association of Science in

---

[2] Pole-Evans was instrumental in starting up a number of important publications. He was involved with the first issue of the *Memoirs of the Botanical Survey of South Africa* in 1919 and subsequently initiated the series *Flowering Plants of South Africa* (1920) and *Bothalia* (1921). The 1929 handbook on *Science in South Africa* contained an early vegetation map in his contribution on 'Vegetation of South Africa'. Later, his vegetation map for South Africa was published in the *Memoirs* (1936). Pole-Evans was involved in a number of major collection expeditions outside South Africa. In 1930 he travelled to Zimbabwe and on to Zambia and Malawi; in 1931 and 1937 he mounted expeditions to various parts of Botswana, including the Okavango delta; and in 1938 he travelled overland to Kenya. Descriptions of these journeys were published after his retirement in the *Botanical Survey Memoirs* (Vols 21, 22, 1948).

1920 he berated officialdom for its lack of attention to the botanical sciences:

> Those of us who toil in the Public Service know to our cost the views held by the average pedantic official, who regards the study of botany as an expensive hobby, to be looked upon with condescending tolerance rather than as a vital necessity of an agricultural country. (Pole-Evans, 1920: 2)

His campaigning paid off. In 1934, the House of Assembly expressed serious concern about the deterioration of natural vegetation and the threats to water resources. Pole-Evans submitted an ambitious proposal for the formation of a Pasture Research and Veld Management Section within the Division of Plant Industry. Following approval, a series of research stations committed to the study of appropriate veld management practices was established in the Transvaal, Natal and Cape provinces. Just prior to his retirement in 1939 he offered a rallying call to his colleagues:

> Grass is the foundation of man's existence in our land as in all others. It is surprising therefore that there should be many who are slow to recognise this and some even loathe to admit it . . . It is my obvious duty again to draw your attention to the fact that large areas of the country which formerly were rich and flourishing pastoral grounds are now wholly depleted of their grazing and are rapidly becoming desert wastes. Nothing but the establishment of well-equipped pasture research stations in these areas can bring any permanent relief and restore health to the land and wealth to the people. (Pole-Evans, 1939a)

Drawing on research carried out in Europe and the United States, Pole-Evans was a strong advocate of rotational grazing systems as the way to manage rangeland effectively. His early advocacy of this approach was to initiate a long history of research on rotational systems and many experiments in grazing schemes in southern Africa. In 1932 he argued that:

> Uniform grazing of the veld should be aimed at. This can only be obtained by a complete system of camping . . . In the US, an ingenious system, known as 'deferred grazing' has been worked out and practised, with highly satisfactory results. There is no reason whatsoever why the same system should not be adopted in the Union. (Pole-Evans, 1932: 917)

Pole-Evans's influence stretched well beyond the borders of South Africa. In the latter part of his career, and following retirement, he travelled widely, advising government officials in Rhodesia and Kenya of imminent environmental disaster and the need for sound rangeland management based on solid research. While not the sole voice making such proclamations across British colonial Africa, Pole-Evans's scientific credentials from South Africa added weight to his authority and experience. For instance, in 1938 he travelled overland to Kenya at the invitation of the Kenyan government. Commenting on Machakos district he noted that the reserve 'was a shambles . . . Sheet erosion and gully erosion were eating the land away in almost every direction. The grass cover had almost entirely disappeared' (Pole-Evans, 1939b: 4). His report provided the basis for attempts at destocking and later the Swynnerton plan for agricultural intensification (Throup, 1987). In 1942 he offered oral evidence to the enquiry of the Rhodesian Natural Resources Board. His view again was that 'the problem . . . is a pastoral problem. You have got to get down to proper grazing management';[3] his recommendations contributed to the draconian Natural Resources Act of 1942 which offered a legislative framework for intervening in resource management issues in the native reserves.

Case study 2: John Phillips
From the mid-1930s, another influential South African ecologist, John Phillips (du Plessis, 1987), was helping to institutionalize the conservationist ideal promulgated by Pole-Evans within the education system. Phillips was born in Grahamstown in 1899 and, following degrees in forestry and botany at Edinburgh University and four years acting as an ecologist in Tanganyika, he took up the post of Professor of Botany at the University of the Witswatersrand in 1931. Phillips published extensively on South African vegetation types and had a special interest in successional dynamics of forest vegetation (e.g. Phillips, 1934). By the mid-1930s he was focusing more on the applied aspects of veld conservation. In 1938 he published an influential article on the 'Deterioriation in the vegetation of the Union of South Africa, and how this might be controlled' (Phillips, 1938). He sums up his assessment of the grassland situation with a call for greater attention to veld management:

> Mismanagement of veld is the main theme in the dirge of deterioriation and depends on the notes of ignorance regarding the nature, behaviour and responses of the grassveld, the lack of rotational grazing and rest systems in grazing. (Phillips, 1938: 479)

He concludes the article with a polemical call to action:

> We do not yet realise, as a nation, that our country's most precious material possessions, the vegetation, its soil, its water, are being taken from us by three national foes - Deterioration, Ignorance and Procrastination. As a united nation, we must fight these and conquer! (Phillips, 1938: 484)

From his position at Wits, Phillips was able to influence a generation of students in the science – and rhetoric – of range management and conservation. Ecological field research was started in 1933 at the

---

[3] Native Enquiry, Zimbabwe National Archives, S988: 7.

Frankenwald Research Station and many key figures, later making their mark in the colonial departments of agriculture in the region, studied there (Roux, 1946). Following World War II, Phillips established BSc courses in Soil Conservation which became the training ground for several hundred men 'who spread the conservation ideal in many parts of Africa and overseas' (du Plessis, 1987: 267).

The emergence of the research establishment
Although the economic pressures of the depression and later the War restricted the growth of scientific research in range management, the setting up of research stations in South Africa and Rhodesia in the early 1930s represented a significant starting point for the establishment of mainstream range science in southern Africa.

The early experiments were simple in design, usually involving only two or three treatments in simple paddock rotation stocking rate trials (e.g. Haylett *et al.*, 1932). Over time, they became more and more elaborate with more and more variables added (e.g. Denny *et al.*, 1977). What did not change, however, were the premises on which these experiments were based. The research designs were based on assumptions derived from the growing school of applied range management in the USA (Stoddart and Smith, 1943), which took as its guiding concept the notion of successional change in vegetation. Clements (1916) was the first ecologist to elaborate a theory of vegetation change that was based on a linear view of change towards an equilibrium point. In the absence of disturbance, this would ultimately lead to a climax vegetation type. However, if grazing intervened a sub-climax vegetation type would result. The trick for range managers adopting this view of vegetation change was to strike a balance between the natural successional trajectory and animal numbers such that optimal returns are achieved.

This approach to range management, where models of beef production were linked to ideas about planned land use, intended to both improve production and prevent land degradation, became firmly entrenched in the thinking of the research establishment in southern Africa and beyond, influencing policy from the 1930s to the present.

Livestock policy in Zimbabwe in historical context

How have the ideas emerging from several generations of rangeland scientists in southern Africa impacted upon range and livestock policy? This section examines the social, political and economic contexts for the evolution of policy in Zimbabwe since the early part of this century, exploring how science and scientists have influenced the debate. Since the 1930s policies have been influenced by two recurrent themes: the need for 'modernization' of the sector and the need to avoid environmental degradation. In the debate on livestock policy, these have been intimately linked, with the mainstream view arguing that the only route to the 'development' of the small-scale livestock sector is through increasing the efficiency of beef production and reducing the damaging consequences of 'traditional' forms of livestock keeping. This, in turn, required a prescribed form of planned and ordered land use.

Forward to modernity

Early colonial administrators in Zimbabwe embarked on an ambitious programme of 'improvement' involving the introduction of grade bulls and pasture grasses usually from outside the country; all part and parcel of the civilizing and modernizing project of the colonial state.

Pedigree bulls were imported into Rhodesia from as early as 1903 (Romyn, 1935), but their performance was often disappointing on the rangelands. This prompted the call for research to increase productivity. With the sponsorship of the Empire Marketing Board and the Rhodesian Government, and paralleling work in other parts of the British Empire at that time, research was initiated at Matopos and Marandellas [now Marondera] in 1931 (Anon, 1931a) which focused on improved beef breeding and management. In parallel, work on pasture improvement was initiated in the early 1930s, with several strands: first, fertilization of natural pastures to increase yields (Husband and Taylor, 1931); second, introduction of pasture grasses (Romyn, 1932) and finally, veld management and rotational grazing (Haylett *et al.*, 1932). This work borrowed directly from research being carried out elsewhere, especially Britain and South Africa where most research officers in Rhodesia had been trained. During the 1920s, research in Britain on grazing management had expanded dramatically[4] and this influenced the uptake in Zimbabwe. It was generally believed that the transfer of the 'civilizing' and 'modern' practices of Europe to Africa would result in improvement. For instance an editorial in the 1931 *Rhodesian Agricultural Journal* commented:

> If improvement is considered essential in such countries as England, how much more is it necessary in Rhodesia, where we are hoping to raise cattle profitability for the English market. Certain essential facts about grazing have already been acquired empirically such as the improvement that followed intensive cultivation for a limited time and the value of paddocking and fertilisation. (Anon, 1931b: 127)

The aim was to improve African cattle breeds and encourage a process of commercialization into the beef economy. The Native Commissioner of Selukwe [Shurugwi] summed up the policy in 1925:

---

[4] Woodham and co-workers, for instance, were often referred to in reports produced in Zimbabwe in the 1930s. Their influential articles in the *Journal of Agricultural Science* (vols 16–20), published from 1926, were important precursors to work carried out in southern Africa.

It is hoped that the natives will realise shortly that it will be to their advantage to raise few graded cattle in preference to keeping a large herd of almost valueless scrub cattle. Until this is realised they cannot hope to compete in the cattle market to their advantage.[5]

The rationale for this policy was clear. Without the native population contributing to grade beef exports, the prospects for the Rhodesian cattle industry in supplying domestic requirements, competing with Argentinian beef, or supplying Britain and South Africa through preferential trade agreements, was limited (Phimister, 1978; Sambrook and Gishorford, 1935).

The research scientists succeeded in achieving remarkable increases in beef output under ranch conditions, as regularly reported in the *Rhodesian Agricultural Journal* during this period. However, this research station success was not repeated in the reserves, where the exotic breeds performed very poorly in the harsh conditions. Despite a programme aimed at distributing grade bulls, the uptake was limited and many died.

This failure is not particularly surprising given that African livestock owners were by and large not beef producers, keeping their cattle for other reasons. The value of a live animal delivering intermediate products such as draft power, milk, manure and so on was far higher than the sale value (Scoones, 1992). Their failure to 'modernize' and sell animals to the formal market – understandable in these terms – frustrated many officials, who sought other explanations for it. Herskovits's 'cattle complex' concept (Herskovits, 1926), in particular, had a great impact on the interpretation of livestock husbandry in the communal areas. The failure to enter formal markets was interpreted in terms of 'culture' and 'tradition' and the need to hold on to animals for 'prestige' and 'status' (Mtetwa, 1978). The standard solution was to 'educate the natives' into modernity, trying to dispel 'backward' notions of keeping animals for security and prestige. However, the propaganda did not work, as noted by the Director of Veterinary Services in his report of 1966:

> The reluctance of Africans to part with animals . . . shows only minor change despite years of propaganda and bitter experience.[6]

Current livestock policy statements echo these concerns. Although there is a growing realization that the communal areas require different forms of support to the commercial sector, the push towards moderniz-ation and commercialization to banish backward, environmentally degrading practices remain dominant influences. For instance, the national livestock policy argues for policies 'to foster proper livestock

---

[5] Native Commissioner's Annual Report, Selukwe (1925), Zimbabwe National Archives.
[6] Director Veterinary Services, Annual Report, Salisbury (1966), Department of Veterinary Services Library, Harare.

management systems in the communal areas' (Zimbabwe Government, 1992: 3). These include 'the improvement of efficiency of livestock production; improvement of marketing services, increasing livestock exports etc' (p.4). This is based on a familiar problem diagnosis that includes 'low offtake rates due to cultural and other factors' (p.11) and 'widespread evidence indicating serious mismanagement of communal area resources as evidenced by high stocking rates and land degradation' (p.13).

The rhetoric of environmental collapse

Environmental concerns have been another undercurrent influencing the implementation of policy and the conduct of scientific investigation since the 1930s. One of the most influential early reports was the Report of the Union Drought Investigation Commission. This was published in South Africa in 1922 and 1923, and received wide coverage elsewhere. The summary published in the *Rhodesian Agricultural Journal* of 1925 offered a particularly florid version of the early environmentalist rhetoric:

> Since the white man has been in South Africa enormous tracts of the country have been entirely or partially denuded of their original vegetation, with the result that rivers, vlei and water holes described by old travellers have dried up or disappeared.. the logical outcome of it all is the Great South African Desert, uninhabitable by man. (Anon, 1925: 771)

By the 1980s the style of the rhetoric had barely changed. Henry Elwell, a respected soils engineer, writing in the *Zimbabwe Science News* in 1983 proclaimed:

> There is absolutely no doubt whatsoever that malnutrition and death through starvation of the Communal land population is inevitable if present rates of soil erosion are allowed to continue. Unfortunately many people still believe that this condition will only be reached some time in the distant future and therefore is not a priority concern. This dangerous fallacy ignores the true facts of the situation and encourages an attitude of complacency. If soil erosion in the communal lands is not checked immediately by a dynamic policy based on reliable technical information, we will witness mass starvation in our life time. (Elwell, 1983: 145)

In the 1920s, officials feared that the dust bowl experience of the United States would be repeated in southern Africa. Alarmist articles were reprinted in the Rhodesian press and professional publications. For instance, a highly dramatic piece by W.C. Lowdermilk of the Soil Conservation Service of the United States Department of Agriculture was reprinted in the *Rhodesian Agricultural Journal* relating the linkages between the collapse of 'great civilisations' and soil erosion (Lowdermilk, 1935). In the same issue, the editorial talks of the 'evil of soil erosion' and sets out the 'ten commandments of good stock farming' to counter it. These apocalyptic and biblical tones set the scene for a general panic about environmental collapse that reverberated across the British colonial territories during the 1930s (Anderson, 1984; Beinart, 1984).

Colonial officials were sent to the United States to examine the situation, returning with alarming stories that only added to the fervour for policy action. While early commentators were cautious, as the colonial state grew in confidence, the calls for direct intervention, particularly in the reserves, grew. Calls for environmental action came particularly from the politicians who saw the policy of Land Apportionment faltering if successful land management was not imposed on the Africans now restricted to the reserve areas. The McIllwaine Commission of Enquiry represented the turning point. The report laid out the dilemmas very clearly:

> The result of the laying waste of large areas of land by wasteful methods of cultivation and the increasing number of livestock has been a cry by the Natives for more and more land. More land would at best be a temporary palliative as present practices, if allowed to continue, would soon render the whole country well-nigh uninhabitable. (Southern Rhodesia, 1939: 14)

The McIllwaine report, supported by the Natural Resources Board Enquiry a few years later, laid the foundations for a series of laws designed to prevent the supposed wanton destruction of the land.[7]

The emotive language and the political interest induced, provided the perfect platform on which the scientific community could contribute technical solutions. These included the construction of contour ridges on agricultural land, the prevention of stream-bank cultivation and the regulation of stock numbers according to defined carrying capacity levels.

Pasture scientists who toured the country to assess the situation reported back in apocalyptic tones. Having previously qualified in South Africa under the tutelage of Professor Phillips (see above), Oliver West was by 1948 the Chief Pasture Research Officer at Matopos. His report on the Belingwe (now Mberengwa) reserve was typical:

> [The grazing areas] have by continuous heavy grazing every growing season, year after year, been pushed to a point from which if relief is not afforded very soon deterioration will become increasingly rapid. The amount of bare ground and erosion already prominent will increase rapidly and the ability of the reserve to turn out cattle in good condition will be seriously reduced . . . The answer is the imposition

---

[7] Building on earlier legislation (e.g. the Water Act of 1928), the Natural Resources Act (1942) provided for a highly interventionist regulation of natural use. The Native Land Husbandry Act (1952) similarly sought compulsion and regulation as the route towards land-use planning and environmental management (Scoones and Matose, 1993).

of a suitable system of deferred rotation grazing . . . but obviously to enable recovery to take place the existing number of stock carried in the reserve should be considerably reduced. (West, 1948)

Scientists like West, were expected to come up with carrying capacity figures for all the native reserves so that appropriate stocking levels could be maintained. These carrying capacity figures were derived from the beef production research trials initiated in the early 1930s and identified stocking rate levels based on an assumed beef production objective which were wholly irrelevant to the African farming setting (Scoones, 1989a). Nevertheless, the policy was backed by law and implemented across the country until 1961. Although officials recognized that the policy was causing untold damage to the pastoral economy and people's livelihoods, the paternalistic attitude among most was that it was 'good for the natives' and essential to save the environment. The Chief Native Commissioner offered his views in the annual report of 1957:

No person likes having to dispose of his property, and least of all the native his cattle. This is one of the many problems, perhaps the most unpleasant of all, that faces the administrative officers of this department. But it has to be done in the interests of the natives and for the future.[8]

However, the opposition to the Native Land Husbandry Act grew and grew during the 1950s and it eventually had to be abandoned due to nationalist opposition. The links between environmental change and land pressure became increasingly highlighted by the nationalist movement during the 1960s. The District Commissioner for Nuanetsi commented in his report of 1964 on the consequences of unequal access to grazing resources:

If physical destocking is out of the question, and it probably is, then I feel we cannot wait until cattle destock themselves by dying through lack of grass. Before this point is reached the tribal trust areas will be dust bowls . . . This 'natural' method of destocking is dangerous from a political angle too. Africans whose cattle are dying through lack of food will decry the existence of grazing on adjoining European ranches and farms.[9]

After independence in 1980 the spectre of destocking reared its head once again. While political constraints meant that destocking could not be countenanced in practice, this did not prevent it being talked about. Indeed, the resurgence of environmental rhetoric in the mid-1980s put the debates started in the 1930s back on the agenda. The need for stock control, better land management and destocking where necessary were

---

[8] Report of the Secretary for Native Affairs (1957), Department of Native Affairs, Zimbabwe National Archives.

[9] District Commissioner's Annual Report, Nuanetsi (1964), Zimbabwe National Archives.

endorsed in the first five-year plan (1986–1990) (Zimbabwe Government, 1986: 27), while the IUCN National Conservation Strategy, fully endorsed by President Mugabe himself, notes:

> In many cases this [environmental degradation] may require that animal numbers are restricted to a stocking rate that does not suppress the perennial grasses in well managed areas and allows the grass cover to recover in degraded areas. (Zimbabwe Government, 1987: 25)

The apparently objectively tested technical solutions offered by scientists, combined with the environmentalist zeal adopted by officialdom, have left limited opportunity for questioning the assumptions behind the doomsday environmental scenario and premises upon which the technical solutions have been built. Political sensitivities both pre- and post-independence have effectively diverted attention from the basic question about the viability of farming livelihoods in the communal areas and the question of access to land has been continuously side-stepped.

Indeed the environmental argument has been used by the commercial farming lobby against the post-independence resettlement programme. In its campaign against the 1993 Land Acquisition Act, the Commercial Farming Union claimed that resettling [African] farmers on [white] commercial land would be tantamount to committing environmental suicide. Commenting in *The Farmer*, Harvey argues:

> In no way can the problems of years of neglect and mismanagement in the existing communal lands be solved by sacrificing further large areas of prime agricultural land to a repetition of this historic process of destruction . . . It must be accepted that the plight and poverty of the communal lands are mainly self-inflicted and that they constitute a national disaster. (Harvey, 1991: 23, quoted by Biot *et al.*, 1992).

Fully aware that the international donors were obsessed with environmental issues around the time of the United Nations Rio Earth Summit, they made much political mileage out of arguments used in different contexts before.

The politics of straight lines
The perceived potential for chaos resulting from environmental destruction resulted in calls for effective planning and control of land use from the 1920s. As the 1939 McIllwaine report noted: 'For the regeneration of the reserves there are two essentials: organisation and control' (Southern Rhodesia, 1939: 57). In particular, the management of grazing through the placement of orderly fence lines was seen as imperative:

> Probably in no other direction could expenditure be undertaken to greater advantage than in dividing pasture lands into suitably

arranged subdivisions.. The benefits to be derived from well placed fencing are so great that no reasonable effort should be spared to bring it within the reach of all pastoralists. (Southern Rhodesia, 1939: 49)

Land-use planning has been used as the centre-piece of environmental and land husbandry policies in Zimbabwe since Alvord's experiments in Shurugwi communal land in the late 1920s. The centralization policy of the 1930s (and subsequently formalized as part of the Native Land Husbandry Act (NLHA) of 1951) was based on the assumption that sustainable forms of agriculture and livestock rearing could only occur with effective planning. Once the land had been divided into viable agricultural units with associated grazing areas, technical solutions to boost production could then be introduced. The grand design for the model African farming landscape was based on linear settlements, dividing an area of individual (male-owned) plots on the upland area and paddocked grazing in the lowlands. Livestock management was to be based on systems of rotational grazing at appropriate stocking rates resulting in high offtake and veld conservation.

The abhorrence of disorder and the apparent chaos of traditional farming systems severely upset officials who were obsessed with the aesthetics of neat and tidy straight lines. Recalling the time prior to centralization in Shurugwi, the Administrative Officer for the NLHA marvelled at the achievements made:

> The whole area had been ruthlessly savaged by the most flagrant abuse of every conceivable concept of land use. The area was hopelessly overstocked, almost completely denuded of vegetation, large scale shifting cultivation left little for grazing and sent the best top soils down the rivers . . . the change over the last 25 years has been little short of phenomenal.[10]

While the scientific community argued for a technical rationale for orderly plans, administrators had additional arguments for straight lines. Even though much of the rhetoric focused on the provision of services to planned settlements and agricultural areas, other motives were apparent, centering on the imperatives of control and surveillance of the colonial state (Robins, 1994). The creation of village settlements ('lines') under the authority of a headman (*sabhuku*) allowed for more effective control in the rural areas. The *sabhuku* (literally the holder of the book), was in charge of the collection of taxes and was responsible for the initiation of development projects being promoted by the state.

Grazing schemes, as part of land-use plans, were promoted vigorously from the 1960s following the collapse of the destocking policy and the NLHA. Drawing on the rotational grazing experiments started in the early 1930s, officials argued for paddocked grazing systems as the model

---

[10] von Memerty, H. (1955), Report on Selukwe Reserve, Unpublished report, Matopos Research Station Library.

for controlled grazing. It was argued that grazing areas could be more efficiently managed if they were consolidated and divided up into blocks divided by fences. Cattle could then be rotated around these camps in an orderly manner. This was seen as a 'positive policy' in contrast to the destocking approach which had been widely resisted.[11] By the early 1970s there were several hundred such schemes in the country, many of which were fenced (Dankwerts, 1973; Froude, 1974). The liberation war interrupted their implementation and many broke down. In many parts of the country, people moved into designated grazing areas and started 'freedom farming', and the linear settlements began to dissolve as people moved to spots of their choosing. At the same time, the grazing scheme fences were cut and redeployed around homes and vegetable gardens. For a period, the politics of the straight line had broken down.

However, straight lines were soon reimposed within post-independence grazing schemes, part of land-use planning exercises with many direct parallels with the previous initiatives (Cousins, 1992; Robins, 1994). While the political rhetoric had changed, the technical rationales had not, and were often justified using the same research; in some cases, the same plans were implemented using the maps and aerial photographs marked up by Land Development Officers in the 1950s. The politics of control in the rural areas were just as relevant in the post-colonial era as they had been previously (Drinkwater, 1989).

The linkages between state-imposed political and administrative control and the science of grazing management are indirect. The fear of environmental collapse and the consequent turmoil that would be unleashed has terrified governments in Zimbabwe since the 1930s. Governments have presided over a highly inequitable pattern of land distribution with a large population crammed into limited areas. The prospect of environmental deterioration leading to the collapse of the agricultural base and to social unrest is, not surprisingly, a cause for political concern. Science has offered both an explanation for the perceived degradation and an apparently simple way of preventing it that involves control of land, livestock and people. The model of land-use planning, stock control and rotational grazing removes the fear of uncertainty and offers the opportunity for the control of the unknown. Science thus has offered and legitimated a solution that has neatly coincided with the political and administrative objectives of governments since the colonial era.

Evidence for environmental calamity

What is the evidence for the doomsday scenario of environmental decline? Commentators from the first decade of the century (e.g. Watt,

---

[11] Secretary for Native Affairs (1957), Department of Native Affairs, Zimbabwe National Archives.

1913) through to the present wave of environmentalist furore (e.g. ENDA/ZERO, 1992; Magadza, 1992) have predicted an imminent collapse. The evidence presented in support of these arguments falls into two types: first, anecdotal reports; and second, quantitative data that has been selectively interpreted.

Anecdotal information, especially if dressed up in the language of disaster, has been very influential. Most of the early commentaries were based on normative interpretations of what the African farming landscape should look like, derived from casual roadside observations. Deviations from this ideal were then deemed to be in need of improvement. This ideal derived from several sources. The European aesthetic ideal of the tidy and ordered landscape was very powerful, as were scientific ideas about optimal farming methods imported from the temperate zones of Europe and North America. These emphasized concepts of carrying capacity, stable grass species composition, mixed farming, and so on; all concepts based on notions of stability and equilibrium, and alien to the highly dynamic, opportunistic approaches to farming found in dryland Zimbabwe.

Quantitative information on land degradation in Zimbabwe is very limited. One of the major sources has been the run-off plots set up by Norman Hudson and colleagues at Henderson Research Station (Hudson, 1957). These data have been successively extrapolated to larger and larger scales in order to estimate the soil loss per hectare, per region and for the whole country. One particularly influential paper used the same data to estimate the economic cost to the country of the loss of key soil nutrients through erosion, assuming that the rates of erosion found in the Henderson plots were found in all areas of the country (Stocking, 1986). The result was dramatic. A total of some Z$2.5 billion was being lost yearly (at 1985 prices) in terms of lost nitrogen and phosphorus alone (Ellwell and Stocking, 1988). But the figures are highly misleading. Not all soil lost from a plot is lost from the system, as any attempt to match dam siltation figures or crop yields with erosion data will show (Biot *et al.*, 1992). Most soil is simply redistributed within the agricultural landscape. The problem is one of scale of analysis; scaling up should only be done when additional data are collected on larger scale processes, lest gross inaccuracies arise (cf. Stocking, this volume).

The danger with quotable figures, especially with dollar signs attached, is that they are used to inform policy. Stocking's figures have been used again and again – in Ministers' speeches (usually with some additional level of exaggeration for good measure), in consultancy reports on environmental issues, in recommendations to policy makers, and in training course materials for environmental managers. Dramatic figures get noted and quoted, and soon they enter the public memory as facts and yet more proof of something that is already believed. It seems that rhetoric and reality have become blurred, as evidence has become less and less necessary to justify action on behalf of the environment.

Challenges to the mainstream

It was not until the 1970s that alternative views of rangeland ecology emerged. A new wave of systems ecologists began to contribute to the understanding of savanna dynamics in southern Africa. Brian Walker and colleagues, first at the University of Rhodesia and later at the University of Witwatersrand, began to elaborate a more dynamic view of ecosystem change, drawing on the concepts of multiple stable states and non-linear dynamics (Walker *et al.*, 1978; Walker and Noy-Meir, 1982; Walker, 1985; Frost *et al.*, 1986).[12]

From the mid-1980s a number of studies began to examine the themes of non-equilibrium dynamics of ecosystems in a variety of contexts. Early work focused on wildlife management, building on a long tradition of population dynamics analysis in this setting (e.g. work in Australia on kangaroo management; Caughley *et al.*, 1987). Challenges to range and livestock management came later. The classic work by Jim Ellis and colleagues in Turkana, Kenya, resulted in a highly influential paper (Ellis and Swift, 1988) challenging the equilibrium view of range management. Similarly in Australia and the United States work on livestock and rangelands (Friedel, 1991; Laycock, 1991) drew on the same ecological themes. The book *Range Ecology at Disequilibrium* provided the first application of these ideas to rangeland systems across Africa (Behnke *et al.*, 1993).

Despite accumulating scientific evidence, this new thinking has been slow to feed through into a critique of livestock development policy in Zimbabwe. However, a number of key individuals, together with an emerging network of research workers drawing on a variety of disciplinary traditions, have begun to make in-roads into the mainstream view.

From the 1970s Allan Savory became a strong proponent of high stocking rates and short duration grazing (Savory, 1988). Using his political position as a leader of the white opposition to Ian Smith's regime, he made strong allies with sections of the white farming community and with post-independence leaders. Through these contacts he was able to experiment on quite a large scale with his unconventional ideas about the role of hoof impact on so-called 'brittle' environments. Although he has failed to make much head-way with the scientific community in the region, who argue that his ideas are inadequately tested, he does have some avid followers among the ranchers and extension workers who have attended his courses in Holistic Resource Management in New Mexico or in Zimbabwe.

---

[12] This work was closely allied to wider changes within ecology and the sciences in general. For instance, theoretical work on population dynamics had been focusing on the implications of non-linear dynamics (May, 1973; 1977). This in turn led to important applications in fisheries, pest management, forest systems, wildlife management and rangelands.

Independence in 1980 offered the obvious chance for a re-examination of the precepts on which previous policy was based. A majority government held power for the first time, and was committed to a transformation of the agricultural sector, putting communal-area farmers at the forefront of development attention. Soon after Independence, Stephen Sandford, a renowned livestock economist and then coordinator of the Overseas Development Institute Pastoral Development Network, carried out a review of the livestock sector on behalf of the new Ministry of Lands, Resettlement and Rural Development. Sandford's report launched a well-argued and devastating attack on previous approaches and argued for a complete overhaul of livestock policy for the communal lands. It de-emphasized the commercialization of beef production as an objective, and instead focused on the opportunistic management of livestock, appropriate to the variable ecologies of the country's dryland zones (Sandford, 1982a). The report was not well received. The Ministry refused to accept his findings and it was again filed away. The scientists who reviewed it were thoroughly schooled in the mainstream view and would not countenance an alternative; particularly one put forward by a social scientist. However, Sandford's report gained a certain notoriety; although never published, it was widely read. It became an essential reference source from then onwards and an important starting point for the many detailed field studies that followed.

A series of studies carried out in the communal areas from the mid-1980s began to amass the more detailed empirical evidence required to challenge the mainstream view. This period saw an interesting constellation of disciplinary perspectives applied to the questions of natural resource management in the communal areas. Ecological analyses investigated the dynamics of savannas under intensive use (e.g. Gambiza, 1987; Scoones, 1990; Campbell *et al.*, 1989); anthropological studies examined people's local resource management practices (Mukamuri, 1988; Elliot, 1989; Wilson, 1990; McGregor, 1991; Matose, 1991) and the impact of development interventions (Drinkwater, 1989; Nhira, 1992; Murombedzi, 1992; Cousins, 1992; Matose and Mukamuri, 1994); historical studies focused on the changing relationship between the peasantry and state intervention (Alexander, 1991; Ranger, 1985); and economic analyses centred on the rationale for communal-area livestock production (GFA, 1987; Scoones, 1992). Most of these studies were carried out by students working and living in communal areas for extended periods. Common to many of them is that they examined actual practice on the ground from an historical perspective, looking at issues at the appropriate scale and drawing policy implications cautiously. All researchers were affiliated with the University of Zimbabwe, most with the Department of Biological Sciences, various parts of the Faculty of Agriculture and the Centre for Applied Social Sciences. The social network amongst this group of researchers was a critical factor in the exchange and develop-ment of ideas. Together, these studies articulate an alternative view that provides a challenge to mainstream policy perspectives on environ-ment and development.

Conclusions: the tenacity of policy myths

Despite this assault on the status quo, the impact of the new thinking on policy remains very modest. Why have a particular set of ideas about grazing management in Zimbabwe (and elsewhere in southern Africa) clung so tenaciously on? Why in the face of empirical challenge has a particular view of environmental management been so dominant over such a long period? Such questions are not easy to answer. This concluding section, however, begins to examine the reasons in the Zimbabwean context.

The fact that rotational grazing is still being promoted in southern Africa – in spite of more than 300 experiments carried out over 50 years or more, which have shown that it does not necessarily result in increased output (Gammon, 1978; O'Connor, 1985) – is testimony to the seductiveness of the simple solution and the fear of the unknown and more complex alternative. Visible, straight, orderly fence lines can be paid for and implemented without worry in a technically efficient manner. It is a clear and simple response to the perception of environmental degradation, resulting from perceived overgrazing and overstocking. Some ideas cling on to public and scientific imaginations long after they are widely discredited by strong empirical counter-claims. Such 'development narratives' (Roe, 1991) usually provide analyses of problems and solutions of particular clarity. They are therefore very appealing to policy-makers as they offer a clear message of what to do and how to spend money. By ignoring the possibility of uncertainty and complexity, science thus offers the chance of dealing with seemingly intractable problems like rangeland degradation. The fear of ignorance is therefore dealt with by suggesting the illusion of control through technical intervention (Wynne, 1992b).

Herders, of course, are continuously innovating and adapting grazing practices in response to the uncertain conditions of the dry rangelands. They must do so in order to survive. However, until recently, this complex, contingent behaviour has gone unrecognized and has not been the subject of concerted scientific enquiry. While herders have been innovating on the rangelands, scientists have persisted with repeating endless variations of their stocking trials and rotational grazing experiments on research stations. While there have been many studies of the complex patterns of indigenous technical knowledge of grazing management and herding practices in pastoral societies (Niamir, 1990), there has so far been a limited articulation of these understandings within livestock and range management science in Africa (Scoones, 1995).

As Wynne notes, referring to the dislocation between the way people respond to uncertain environments and the way science attempts to deal with uncertainty in terms of technical interventions or public policy recommendations:

> Ordinary social life, which often takes contingency and uncertainty as normal and adaptation to uncontrolled factors as a routine necessity, is in fundamental tension with the basic culture of science, which is premised on assumptions of manipulability and control. It follows that scientific sources of advice may tend generally to compare unfavourably with informal sources in terms of the flexibility and responsiveness to people's needs. (Wynne 1992b: 120)

Political explanations for the persistence of particular ideas can also be sought. The land-use planning exercises initiated in the 1930s were justified in terms of 'scientifically rational' and 'environmentally sustainable' land use, yet had other aims to do with tax collection, political control and so on. Since independence, attempts to suppress rebellion in Matabeleland can be linked to livestock and range management 'development' projects that provided for consolidated settlements with access roads allowing surveillance and control by the army (Robins, 1994). On this reading, the discourse of development intervention centres around the exertion of power by a bureaucratic state apparatus, and the depoliticized technical language of environment and development provides a useful means to justify intervention in the name of technical progress, while at the same time fulfilling political aims of control (cf. Ferguson, 1990).

However, the repeated attempts to control and manipulate peasant livestock management strategies in Zimbabwe have dramatically failed. The politics of domination and control are not all-powerful. The agency of local actors in mounting resistance has been a recurrent theme (cf. Scott, 1985). Whether resistance has taken a passive form of non-compliance (as in the beef breeding attempts) or more active forms of opposition (as in the case of destocking), people have preferred to follow their own informal and flexible alternatives in order to survive. Resistance to imposed technical solutions to problems diagnosed from outside has characterized the history of livestock policy interventions in Zimbabwe. Many herders understand the principles of paddocking and rotational grazing, as they have seen such systems operate on nearby commercial farms. It is not ignorance that engenders this resistance, but a more fundamental disquiet about the technical rationale for the suggested solution under local circumstances and a suspicion about ulterior motives. Many grazing schemes have faltered during their implementation due to social and institutional factors that create an atmosphere of distrust between local people and the extension agents, the purveyors of the scientific solution. The top-down approach to implementation limits the possibilities of exchanging perspectives and negotiating outcomes between local herders and external agents. The result is the emergence of forms of resistance that are actively pursued, but perceived by outsiders to represent ignorance of the 'correct' solution, implying that people require education and persuasion (Scoones and Cousins, 1994).

The persistent perception that ignorance is at the root of non-compliance does not lead to the questioning of the assumptions behind the intervention and a re-examination of its scientific premises. Once proven on the research station and tested on commercial farms, the solution remains sacrosanct and the myth continues to be propagated through the educational institutions and in the scientific establishment.

Censorship of new ideas is often highly effective. New ideas may be dismissed as maverick or unsubstantiated, challenges will not find their way onto the pages of mainstream journals controlled by the scientific establishment, or reports may be boycotted. Forms of censorship have been prevalent in Zimbabwe. The key reports of Theisen and Marasha (1974) and Sandford were never published, while the work of Savory is simply dismissed as unscientific. By contrast, the local journals are full of papers on conventional grazing experiments; all marginal adjustments of previously published data.

Over the last decade, the challenges to mainstream range management have begun to coalesce around an increasingly coherent reinterpretation of range ecology dynamics together with a reassessment of the direction for livestock development. The village-based studies informing these have formed a growing body of empirical evidence that is increasingly being used as source material for policy work and further studies.

While these new perspectives on resource management are increasingly being taught in the University of Zimbabwe and other key educational institutions in the region, and students are beginning to enter jobs in the scientific research and extension establishment, the impact on development practice on the ground still remains limited. Although individuals may have shifted in their perspective (and there have been some notable conversions), institutions are slower to move (cf. Kuhn, 1962). The inertia of large bureaucracies is notorious. The opportunity for new ideas to spread is highly constrained, especially if they have the potential fundamentally to shake up the rationale for the existence of the organization. The agricultural extension organization in Zimbabwe, Agritex, is a good example. There are a number of key individuals in the service who are proponents of a radical reconception of livestock management approaches yet, despite being in relatively senior positions, they have to convince several thousand extension workers in the rural areas to move with them. At the same time as they are advocating a new approach, the training colleges and the in-service training unit continue to use course materials that resemble things that were being taught in the 1930s. Ideas move on, but institutions and organizations so often stand still (Chambers, 1993).

The future remains uncertain. Will the implications of the new range ecology, centred around flexibility and opportunism, be acceptable to the political objectives of the state and development agencies? Will the imperatives of external control and regulation be reimposed? Will the popular grip of environmental calamity maintain its hold? Or will scientists and development practitioners, convinced of an alternative to the mainstream, be able to strike an alliance with the herders in the communal areas who have been practising opportunistic management all along?

3

Soil Erosion Animals & Pasture over the Longer Term

Environmental Destruction in Southern Africa

WILLIAM BEINART

Introduction[1]

A great deal of the literature on South Africa which explicitly addresses environmental issues paints a picture of decay and degradation over the long term. One of the most persistent concerns, and an important focus for commentators, has been soil erosion (Beinart, 1984). In a country where livestock were important both for white and black farmers, overstocking and overgrazing have frequently been cited as major causes of denudation, desiccation and erosion. The state of the settler stock farms and semi-arid areas more generally have been the trigger for broader debates and discourses about ecological decay in the region. This chapter examines ideas about degradation on white-owned stock farms in the lower rainfall grazing regions of South Africa – more than half the area of the country – in the light of some fascinating recent research in botanical history.

Hoffmann and Cowling (1990) are uneasy about the scenario so frequently painted of continuous degradation in natural veld pastures in the Karoo and its better-watered grassland peripheries. To some degree their findings, challenging received wisdom, echo an earlier but minority botanical viewpoint. They tend to explain their less alarmist approach by emphasizing that the vegetation of this area has been subject to fluctuation shaped largely by natural causes. This chapter offers some other tentative explanations as to why, from a historian's point of view, their botanical position may have some foundation.

My approach here is cautious, not only because it is difficult for a non-specialist to interpret complex and sometimes conflicting scientific data, but also because there remains much historical research to be

[1] I would like to thank Timm Hoffmann for his comments, for keeping me up to date with botanical debates and for sharing information on John Acocks, whose career he is researching. A version appears under the title 'Environmental degradation in sheep farming areas of South Africa: soil erosion, animals and pasture over the longer term', in G. Chapman and T. Driver (eds) *Timescales of Environmental Change* (Routledge, 1996).

done. Nevertheless, the argument seems consistent with two very general points which do not in themselves depend upon the material analysed here. First, while there is no doubt that growing population, stock numbers, commercialization and apartheid have contributed to environmental problems in southern Africa, environmental historians are on shaky ground if they stalk the past only with the limiting concepts of decay and degradation. Environmental change should be examined in a less linear manner; deployment of the concept of transformation, rather than just degradation, might help to shift the emphasis of debate. Second, it may be unwise to use arguments, canvassed in recent years, about the environmental destructiveness of settler capitalist agriculture in order to justify totally new ownership regimes in South Africa.

The expansion of grazing

South Africa's Karoo, together with its better-watered eastern fringes and the southern highveld grasslands of the Orange Free State (OFS), has been browsed and grazed for a very long time by a large variety of indigenous wildlife species. Before settlers conquered and took control of the region in the eighteenth and early nineteenth centuries, parts had been used for many centuries by the flocks and herds of the Khoikhoi. African people settled on the East Coast also used the grassland fringes seasonally. It is therefore difficult to argue for a pristine period when vegetation existed in a natural state, unaffected by animals and people.

The initial expansion of the settler grazing frontier (c. 1700–1830) was driven by the market for meat. Settlers ran cattle and the fat-tailed hairy sheep which they had adopted from the Khoikhoi, but they used the veld more intensively. The myth of the subsistence-oriented South African trekboer has long been disputed by historians; as on other colonial frontiers, settlers' mode of life could be rapacious. One indication of this is that as early as the 1830s, much of the wildlife had been shot out of the more heavily stocked districts both to secure meat supplies and to reduce competition for grazing. Two species, the quagga and bluebuck, were extinct in the wild. Imported technology (notably guns and horses), together with social disruption in this early colonial period, simultaneously brought new communities of Griqua (regrouping slaves and Khoikhoi) and Africans into the semi-arid interior.

From the 1830s, woolled sheep, mainly Merino, were absorbed into a funnel of land between the 250 mm and 600 mm rainfall lines, spilling over into the wetter East Coast districts. By the second half of the nineteenth century these thinly populated districts had become major growth areas in the Cape Colony and OFS. Although the value of wool exports was overtaken by diamonds and then dwarfed by gold, wool remained the most important agricultural export well into the twentieth century. Figures show extraordinary growth of sheep holding in South

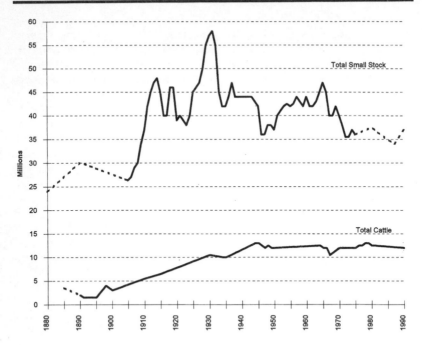

Figure 3.1: Livestock totals, South Africa, 1890–1990 (million head)

Note: This graph represents only approximate numbers. It attempts to include all the stock in
South Africa, whether owned by black or white farmers. At least until the 1950s, many of
the stock on white-owned farms belonged to black tenants and workers. A total figure also
allows for continuity in boundaries: the reserve or homeland areas have approximately
doubled in size since the early twentieth century and the white-owned farmlands have
declined in extent. The original figures in agricultural and other censuses are somewhat
suspect, and numbers are especially weak since the 1970s from which time homeland
statistics are decreasingly reported in national figures.

Africa. It is likely that there were less than four million sheep in 1840,
rising to ten million by 1865, and close to 25 million by the 1891
census; additionally, fat-tailed sheep, angoras and other goats were
numerous in particular districts. Numbers declined during the South
African War (1899–1902) but then grew rapidly to peak in 1913 at about
28 million woolled sheep and 47 million small stock. In 1931, they
reached an all-time high of about 45 million woolled sheep and 58
million small stock (see Figure 3.1).

While these were largely sheep districts, some also carried cattle.
Cattle numbers fluctuated more wildly around the turn of the century
because of diseases such as rinderpest (1896–7) and East Coast fever
(1904–13), as well as the South African War. Nevertheless, the overall
trend was rapidly upwards from under four million in 1904 to about
eight million in 1920 and over 12 million by 1939. The most spectacular

growth was in the OFS and Transvaal, but even in the Cape numbers doubled over less than four decades. Whereas the great majority of sheep were owned by white farmers, nearly half the cattle in the country were, until 1930, owned by black farmers and tenants.

Various factors facilitated this expansion. One was the control of new animal diseases such as scab, redwater, rinderpest, East Coast fever and heartwater – a major preoccupation for colonial agriculture departments. Near-universal dipping of sheep and cattle was achieved by about 1920. Water had also been a major constraint on pastoral farming, with the original location, size and layout of farms shaped not least by the availability of natural water sources – springs, vleis and streams. Dam building was one alternative means of controlling supplies and from the mid-nineteenth century shallow earth farm dams began to pepper the countryside. Boreholes – drilled by the Department of Irrigation as well as private firms – increasingly became an alternative as technology improved in the early twentieth century. They tended to be more reliable in that they tapped underground sources which were relatively constant. Metal windmills sprouted across the dry regions of the country. South Africa did not quite become a new 'hydraulic society' after Worster's (1985) model of the American West, but rather, in stock-keeping country, a borehole society. A third factor was the reduction of predators of small stock, especially jackals and caracals. In the early decades of the century, farmers and the state acted in concert to eliminate these 'vermin', and hundreds of thousands were killed in response to the high bounty payments offered (Beinart, 1993).

The early establishment of the veld degradation orthodoxy

The cost of such rapid expansion of animal numbers, observers increasingly argued, was in the state of the veld. There is a long record of colonial comment on its deterioration, dating back to the period of Dutch East Indian Company rule before 1806. Some of these accounts were collated by T.D. Hall, perhaps the first to attempt a systematic history of pastures, in the 1930s and 1940s. An English-speaking South African (b. 1890), brought up in the OFS and trained in the USA, he was agricultural adviser to the leading fertilizer company in the country. After extensive research into soil chemistry and veld management, he became a strong conservationist and this led him to explore past commentaries. He found officials near Cape Town commenting in 1751 on 'the disappearance of grass and the springing up of small bushy plants in its stead' (Hall, 1934: 66). In 1775 the Swedish traveller Sparrman noted that

> in consequence of the fields being thus continually grazed off and the great increase of the cattle feeding on them, the grasses and herbs

which these animals most covet are prevented continually . . . from thriving and taking root, while on the contrary the rhinoceros bush which the cattle always pass by . . . is suffered to take root free and unmolested. (Hall, 1934: 66)

Such observations were in turn linked to wider developments in European thought concerning the disturbance of natural resources by agricultural development and intensification (Glacken, 1967). Grove (1995) has illustrated the extent of Dutch environmental concern about their new colonies as early as the seventeenth century. He suggests that colonial experiences, especially in transforming the ecology of small islands, fed back as a major strand into developing western environmental thought. By the mid-nineteenth century, following the burgeoning of enlightenment science, many of the linkages which currently inform environmental thinking had been made, in particular those between denudation and soil erosion. 'Desiccationist' ideas were influential. John Croumbie Brown, an innovative Scots colonial botanist at the Cape in the 1860s, wrote extensively and alarmingly on these issues and proposed far-ranging solutions (Grove, 1987). By 1873, a short report by John Shaw to the Royal Society outlined the key points which were to become commonplace over the next century (Shaw, 1873). After sketching the distribution of plants in South Africa, the author went on to specify the character of the prairie-like midlands of the Cape with their luxuriant grass and vegetation:

Since sheep have been introduced the grass has fast disappeared, the ground (by the hurried march of sheep for food amongst a scattered bush) has become beaten and hardened, and the seasonable rains which do come are accordingly allowed to run off the surface without soaking into the ground to the extent formerly the case. The country is thus drying up, the fountains becoming smaller and smaller, and the prospect is very clear that the midland regions will turn into a semi-desert. Indeed the plants of the singular regions known as the Karoo, in the south-west of the Cape. . .are travelling northwards rapidly and occupying this now similar dry tract of country. The herbage is essentially a Karoo one already. (Shaw, 1873: 105)

Between the 1910s and 1930s, when the growth of stock numbers was particularly alarming, T.D. Hall was only one of a number of intellectuals addressing the veld degradation problem. A Senate Select committee of 1914 produced an influential report on 'Droughts, rainfall and soil erosion'. Afrikaner intellectuals, some of whom had trained in the USA, were at the forefront of concern. Enslin, the Chief Inspector of Sheep in the Department of Agriculture, laid particular stress on the damage done to the veld by 'tramping' – both the seasonal treks of animals in long-distance transhumance, and daily movements to widely dispersed water sources on farms. Seasonal burning of veld attracted widespread criticism. The Drought Commission report of 1922–23, which concentrated on stock farming, took extensive evidence and brought together many previously articulated anxieties in powerful and persuasive language. Overstocking, denudation and drought, threatened a 'newly-created South African desert'. The old-established practice of 'kraaling' sheep at night, or bringing them back to a central byre in order to protect them from theft and predators, was particularly criticized (Union of South Africa, 1923).

The elaboration and persistence of the veld decline orthodoxy

By the interwar years, then, environmental alarm, and the understanding that South African pastures were overstocked and had been in continual decline over a long period, had become firmly established. The debate had shifting emphases, a detailed history of which has yet to be constructed. While earlier commentators seem to have emphasized changes in vegetation, climate change, overstocking and soil erosion later became more prominent concerns. And whereas anxiety was first articulated about extensive white-owned farms it was later extended to African reserves. The Native Economic Commission report of 1932 argued urgently that environmental degradation in the reserves implied rapid African urbanization which would cut across the state's segregationist policies. Nevertheless, the basic points about veld decline became remarkably well-established and have permeated a great deal of literature ever since.

Following the Second World War, significant legislative action was taken both in white-owned and black-occupied districts. Soil conservation became a part of post-War reconstruction plans. A further enquiry by the ominously-named Desert Encroachment Committee (1948–51) aimed to examine more definitively whether 'man-made dessication had altered the natural condition of the veld to such an extent that the climate itself had in turn been affected' (Union of South Africa, 1951: 2). The relationship between rainfall and denudation, extensively researched in Europe and the USA, had long been debated in South Africa, with Croumbie Brown in the nineteenth century and many since arguing for a strong correlation. As the Committee reported, 'at all times there appears to have been a fairly widespread belief that climatic conditions were slowly, but progressively deteriorating' (Union of South Africa, 1951: 1). However, as in the case of the Drought Commission 30 years earlier, the report was sceptical about any long-term decline in rainfall. Records suggested that there had been a rainfall peak around 1890, and a gradual, irregular decline until the droughts of the early 1930s. But subsequent rainfall levels had been relatively high, not least immediately before the Committee reported. The report thus accepted the probability of long-term rainfall cycles, while acknowledging that these were difficult to detect with the available evidence.[2]

On the issue of veld deterioration, however, the committee expressed no such scepticism. Most of the evidence was again taken in and around the Karoo. The report was also strongly influenced by the work of John Acocks, an ecologist who had worked for the Department of Agriculture since 1935 and conducted extensive fieldwork. He completed his own national survey, *Veld Types of South Africa*, first published in 1953 (Acocks, 1975), and also served on the Desert Encroachment Committee. Like the Drought Commission, Acocks's work helped to fix the language of debate. It remains the most widely quoted work on pastures and veld types and, although it did not devote an extensive section to pasture history, the work's authority was such that Acocks's observations have become commonplace.

Acocks noted the general desiccation of the Karoo. Indeed, in its drier western parts it had turned into 'near desert in the sense that soil erosion is universal and that there is no longer a permanent, unbroken vegetation cover, and only rarely a temporary cover' (1975: 8). Most strikingly, he reiterated the points made by Shaw, and others since, that Karoo vegetation had moved eastwards, in places up to 250 km, to replace productive, 'sweet' grassveld. There was also a slower, but accelerating northward movement. Elsewhere, sour or rank grassveld, less suitable for year-round grazing, had spread at the cost of mixed and sweeter velds, especially the climax grasses such as *Themeda triandra* or *rooigras*.[3] In short, Acocks claimed that there was 'widespread deterioration in all veld types over the last 500 years' (1975: 78). He was particularly concerned about the development of what he called the False Upper Karoo:

> The development of this veld type constitutes the most spectacular of all the changes in the vegetation of South Africa. The conversion of 32,200 square km of grassveld into eroded Karoo can only be regarded as a national disaster (1975: 78).

The Desert Encroachment Committee incorporated these ideas and reiterated a view often expressed in the inter-war years that the state should take a lead in resolving the crisis, since 'it is quite clear that farmers do not know what is best' (Union of South Africa, 1951: 7). Members also examined research on the siltation of dams and, less conclusively, on the possible depletion of underground water as a result of the rapidly increasing number of boreholes.

It would not be too much to say that the determination of correct systems of veld management combined with the optimum stocking

---

[2] The evidence of such a rainfall cycle is now more widely accepted. Furthermore, there is also evidence of some form of shorter-term (7–8 year) cycle on the East Coast, albeit irregular.

[3] Redgrass – pastures in which it dominates have a deep red-brown colour in winter.

rate for every veld, soil and climatic zone would be one of the greatest contributions that this generation could make to the future welfare of agriculture in South Africa. (Union of South Africa, 1951: 7)

In language very reminiscent of the Drought Commission report and of American inter-war commentators such Sears (1935), the Committee concluded that 'the very existence of stock farming is at stake because, unless we can succeed in arresting further deterioration of the veld, it is doomed' (Union of South Africa, 1951: 7).

Acocks spent some of his career based at Grootfontein, the major agricultural research station and training college in the Karoo, founded in 1918 at Middelburg. There, Roux developed further research over a long period from the 1960s, confirming many of the tendencies highlighted by Acocks (Roux *et al.*, 1981, 1983). He suggested a five-phase progression of deterioration from pristine veld to desert. Much of the Karoo was in the third phase; woody invader plants were still spreading 'at an alarming rate'.

Veld decline and anti-apartheid debates

It is important to note that most of the authors of this twentieth-century material – and a good deal more could be cited – were progressive farmers, scientists and/or government officials, generally not antagonistic to the broader segregationist policies being pursued by South Africa's successive governments. Yet this powerful and well-established technical thinking was readily incorporated into a far more critical anti-apartheid literature on the rural areas. Aside from the fact that physical evidence of decay could be observed in the form of *dongas* (erosion gulleys) for instance, it is not difficult to understand why. On the one hand, ideas about degradation provided an argument with which to castigate the greed of white farmers who were at the core of support for apartheid and were, together with the state, seen as responsible for severe rural dislocation through forced removals. On the other hand, they helped to illustrate the iniquities of the homeland system in which blacks were restricted, as independent occupiers, to a very limited proportion of the country's land. Whereas the Native Economic Commission had blamed African culture and attitudes for ecological degradation in the reserves, it was not difficult to invert the argument and pin the responsibility on the restrictive policies of apartheid.

Many of the more general publications dealing with apartheid and environmental decay correctly illustrate the way in which severe environmental ills have been concentrated in the homelands. Yet a number also summarize key points in regard to white-owned farms. *Uprooting Poverty* (Wilson and Ramphele, 1989) and *Restoring the Land* (Ramphele, 1991), two important books by anti-apartheid social scientists which deal with poverty, ecology and post-apartheid reconstruction, are good examples. Despite government measures, they argue, 'the rape of the soil has continued'. Wilson quotes the estimate that 400 million tonnes of topsoil were being lost annually in the 1960s. Overstocking on sheep farms in the Karoo and Cape Midlands was around 30 per cent and desert advancing at about 2.6 km per year. He reiterates Acocks's view that over the centuries sweetveld had become scrub and that by 2050 the sheep farming areas would be desert. 'The combination of soil erosion and deteriorating vegetation is held by some to be far and away the most serious of the many problems facing South Africa' (Wilson, in Ramphele (ed.), 1991: 30).

Durning's hardhitting pamphlet, *Apartheid's Environmental Toll*, reiterates some of these points (Durning, 1990). He actually conjures up the image of the 'southwestern deserts . . . marching to Pretoria, expanding across two and a half kilometers of exhausted pastures a year' (1990: 6).[4] Huntley, a leading South African environmental scientist, and his co-authors see the 1980s as 'the decade the environment hit back . . . for over a century of careless environmental management' (Huntley *et al.*, 1989). They do not accept a doomsday scenario where resources are finite and population growth of necessity a route to disaster. Nevertheless they do accept the idea of continuous decline manifest in the unparalleled diversity and seriousness of 'natural' disasters – including floods, droughts, fire and locusts – which they argue were worse even than in the 1930s. 'We have had to pay the price for over a century of careless environmental management and South Africa's unique experiment in social engineering' (*ibid.*: 37). They similarly estimate that annual losses of soil are 300–400 million tonnes, nearly three tonnes per hectare, and that the Karoo is still overstocked by almost 50 per cent (*ibid.*: 39). Overall, they argue, three million hectares have become unusable because of erosion. They also express concern, manifest at least since the mid-nineteenth century, about declining biodiversity.

The botanist Cowling echoes these concerns with biodiversity, noting that a particularly large number of species are at risk because of the unusually diverse and unique nature of Cape plants and animals. He argues more generally that:

> The most important cost associated with agricultural development in Southern Africa has been the degradation of natural resources and ecosystems . . . The erosion of topsoil, which is practically a non-renewable resource is unacceptably high . . . Most of the natural grazing land is seriously overstocked and as much as 60 per cent of the veld is currently in poor condition. (1991: 15–16)

---

[4] In this echo of a South African War song, Durning omits to mention that the Karoo would have to cross over 500 km of the better-watered highveld crop belt if it were really to reach Pretoria.

Most of these authors are reluctant to blame capitalism itself for veld degradation. Nevertheless, Cowling, approaching the issues from an ecological viewpoint, comes up with some very radical suggestions. Because it is wrong, he argues, to confine stock to a small area in semi-arid zones 'an ecologically appropriate intervention in the Karoo might involve the removal of barriers to stock migration and nationalisation of the herd' (1991: 17). Taken to its logical conclusion, this analysis might suggest that 'ecological conditions would favour communal land tenure and nomadic pastoralism' (*ibid.*).

Challenges to the degradation view

The idea of continuous veld degradation cannot be seen as an unquestioned nor unfounded 'conventional wisdom'. In the first place, many of the protagonists of this view regarded themselves as somewhat unconventional in the face of the weight of a powerful farming lobby or, later, an uncaring government bent on its programme of massive social engineering. Second, some of the observations about ecological decay undoubtedly ring true. Certainly for the period up to the early 1930s, the growth of stock numbers, especially given the methods of stock farming practised, must have had major and far-reaching ecological effects.[5]

However, important questions can be raised about the lineality and character of change, and the concept of degradation itself. It is striking that a continual build-up of stock numbers was possible for many decades after alarmist reports in the nineteenth century. And after apparently devastating levels of stocking around 1930–31, and calamitous losses in the early 1930s, numbers of both sheep and cattle rose quite sharply following better rainfall in the mid-1930s. Similarly, after a further fall in stock numbers and a further 20 years of reported desertification and the dire warnings of Acocks and the Desert Encroachment Committee, small stock numbers again rose sharply in the 1950s (see Figure 3.1).

One reason for the resilience of national small-stock numbers up to the early 1960s may of course be that increases took place in newly-stocked districts, while areas which had long been heavily stocked in the nineteenth century experienced more severe losses. It is true that sheep were increasingly incorporated into the agricultural systems of parts of Natal, the northern OFS and the Transvaal in the early decades of the twentieth century. And according to the Drought Commission of

---

[5] While it is difficult to make cross-national comparisons because of the variety of ecological conditions, South Africa does appear to have been comparatively heavily stocked at this time. Only Australia, India, the Soviet Union, the USA, and possibly China and Argentina – all far bigger countries – had more woolled sheep; countries like New Zealand and the UK had far higher numbers per hectare but are of course much wetter.

1923, degradation was already leading to a decline in numbers in the old sheep districts of the Karoo and midland Cape, some of which had been well-stocked for over a century. A definitive response to this contention would require careful district by district analysis of stock numbers over a long period. But although the statistics require further investigation, preliminary analysis suggests that it does not hold in any simple sense for the first half of this century. For example, sheep numbers increased quickly in the Cape Province during the 1920s, immediately after the Drought Commission reported that they were undergoing a long-term decline. Moreover losses both in sheep numbers and in wool production in the great 1930s drought were proportionately worse in the OFS, Natal and Transvaal than the Cape – although the Cape had been stocked for a longer period.[6]

Furthermore, different approaches to the question of the relationship between veld status and stock numbers are possible. These do not necessarily contest the view of environmental degradation in some periods, but they do call into question the idea of continuous long-term decline. One of the botanists serving on the Desert Encroachment Committee was C.E. Tidmarsh who also worked at Grootfontein College of Agriculture. To a greater extent than other experts, he was cautious about a linear view of decline. Reporting on some years of veld experiments, Tidmarsh agreed that knowledge was still inadequate, but noted nevertheless that:

> The results of the past seventeen years of research at the College have shown clearly that the amount of natural vegetation that can be maintained per *morgen*[7] of land, is controlled more by the available moisture supply than by the grazing treatment to which the veld may be subjected, and that in the extensive flats of the Mixed Karoo, the quantity of vegetation growing at present on the soil is, with the exception of local areas of denuded soil, in approximate equilibrium with the available moisture supply, and that, without increasing the latter, it is virtually impossible to increase the natural cover of the soil by any measure of grazing control, including complete protection (Tidmarsh, 1952: 1).

Even more controversially, Tidmarsh argued that while extremely heavy stocking rates could change the composition and quality of the veld, it was very difficult to produce a lasting impact. 'Within limits, the composition of the veld appears, thus, to be more a function of the interaction of the soil type, moisture supply, and climate, than of the grazing treatment' (*ibid*.). Noting similar results in the USA, he tentatively suggested that continuous grazing at a moderate stocking rate was not necessarily harmful. It is interesting that Tidmarsh's contention

---

[6] Numbers have been taken from the general censuses (up to 1911), the reports of the Chief Inspector of Sheep, and the annual Agricultural Census (from 1919).

[7] The *morgen* was the unit of land measurement in South Africa before metric units were introduced. One morgen = 2.1 acres = 0.86 ha.

does not contradict the findings of an American agricultural economist who reported on South Africa during the 1930s depression (Taylor, 1935).[8] He calculated that the number of sheep per acre in any district correlated strongly with average rainfall (cf. Behnke and Scoones, 1993; Brockington and Homewood, this volume). In some ways these views also echo the arguments of the American pasture historian J.C. Malin who de-emphasized the role of people, and particularly European settlers, in shaping pasture histories over the long term in the USA (Worster, 1979).

In post-depression USA, where soil conservation provided a major justification for state intervention, Malin's views were associated with an anti-federal and rather conservative anti-interventionist political viewpoint. For Tidmarsh, Acocks and most other South African experts, by contrast, the role of the state was unproblematic – the problem was more how to intervene effectively. Tidmarsh did recognize that there were limits to safe stocking. While he and Acocks differed as to the long-term effects of pasture use, they both agreed that selective grazing of more palatable species could have serious short-term consequences. It had long been argued in South Africa that grazing in fenced paddocks, rotated through the year, increased the carrying capacity of the veld. They refined experiments on systems of rotation and generally supported more intensive use of smaller paddocks, rotated frequently in order to minimize selective grazing.

Echoes of Tidmarsh's approach to long-term vegetation change can be found in new botanical work by Hoffmann, who is cautiously beginning to contest the view that Karoo vegetation has been extensively altered and is spreading (Hoffmann and Cowling, 1990).[9] Together with Cowling, he has assembled a formidable range of historical sources on the appearance of the Karoo before the period of intensive farming of woolled sheep. From early written accounts they argue for the eastern Karoo that 'although there are some references to a grass-dominated landscape, even the earliest accounts suggest that, at least in places, dwarf karroid shrubs were dominant' (Hoffmann and Cowling, 1990: 289).

Second, and more dramatically, they have been involved in systematic photographic research. In view of the long concern about the state of the veld, considerable photographic records taken by botanists survive. These include a series by Pole-Evans, the Cambridge-trained botanist who came to work as a government scientist in South Africa in 1905 and emerged as one of the most influential figures in stimulating research on grazing problems and conservation more generally

---

[8] The report was largely written in order to assess South Africa's capacity to supply international commodity markets and compete with US produce.

[9] Hoffmann himself remains cautious about this argument because the article has triggered a range of responses, some mentioned below, and the scientific evidence is still being accumulated and debated (personal communication).

(Scoones, this volume). Matched photographs taken in 1989 suggest that the state of the veld has improved considerably since 1917–25 when the earlier photographs were taken.

Third, Hoffmann and Cowling resurveyed 11 sites in the Karoo and southern OFS which had previously been investigated by Roux in 1961–3. Roux, who became head of Grootfontein, participated in the sample surveys in 1989 so that better comparability could be achieved. They found that 'All sites showed an increase in total percentage canopy spread cover from 1961–63 to 1989, attributed chiefly to an increase in the cover of grasses' (Hoffmann and Cowling, 1990: 290). Some sites showed a decline in shrub cover.

On the basis of these findings the authors suggest there is no strong evidence for an expanding Karoo. Rather they see the possibility of short-term changes in response to rainfall. Nor do they consider the prevalence of grass cover recorded in 1989 as evidence of a reversal of processes noted earlier, so that grasses are now invading the Karoo. This phenomenon is also more likely to be part of shorter-term cyclical change, possibly linked to higher rainfall in the mid-term to late-1980s. 'Except for a very general understanding, we do not know what influence grazing has on these processes' (Hoffmann and Cowling, 1990: 292).

Veld changes and historical evidence of grazing pressure

While it is difficult for a historian untrained in botany to evaluate these various studies, it is possible to compare them with other historical evidence, notably long-term stocking rates and other factors governing the intensity of veld use. Figure 3.1 gives an approximate picture of national trends in stock numbers. It suggests that the veld photographs for the period 1917 to 1925 were taken after a decade of spectacular growth in small-stock numbers – from about 26 million to 47 million – and when numbers reached their early peaks. Cattle numbers were also increasing rapidly, although not necessarily in the sheep districts. Given that stock were grazed almost entirely on the open veld at the time, and that paddocks were not universal, it is unsurprising that the pictures present a denuded landscape. Surveys carried out in 1962–63 by Roux and his colleagues similarly took place at a time of high small-stock numbers – near to their post-Second World War peak – after more than a decade of sustained increase.

By contrast, 1989 was at the end of a period in which small-stock numbers had been low for over 15 years. Indeed from the early 1970s until the late 1980s, small-stock numbers in the country as a whole were lower than at any time since the first decade of the century, and cattle numbers were also fairly stable. If the figures for white-owned sheep alone are taken – accounting for the bulk of farm animals in the districts under discussion – they show an even more significant fall in the 1980s. Although they rose a little in the late 1980s, they have remained low since then because of the collapse of the wool price and serious drought in the early 1990s. If this evidence of small-stock numbers is put together with the claims of Tidmarsh and others about the capacity for veld to recover under favourable circumstances, it is not entirely surprising that Hoffmann and Cowling's evidence showed improvements.

Two further related issues should be considered: first, whether there are other factors which might have led to veld improvement; and second, whether stock numbers have remained low simply because the veld could carry no more. Absolute numbers are by no means the only factor which might affect pastures. On the one hand, the size and feeding capacity of sheep has probably increased in the period under discussion; 35 million small stock in the 1980s might well eat as much as 45 million in the 1910s. On the other, there has been widespread investment in grazing management. The Drought Commission of 1922–3 predicted that the carrying capacity of veld could increase by 33 to 75 per cent by the use of camps and that the output of wool could double. Internal divisions on sheep farms go back a long way. Despite the size of the farms, dry stone walls were built in the nineteenth century and these can still be seen in some of the older, and formerly wealthier sheep districts. There was certainly no shortage of stones in the Karoo. Barbed wire replaced stone walls from the 1880s, and was increasingly important as a means of controlling jackals. And the expense of fencing was diminished by state subsidies in the early decades of this century.

T.D. Hall noted that Graaff-Reinet farmers, in the heart of the old sheep country, had generally fenced by the time of the 1930s droughts. Although they farmed in an area with only about 350 mm of rainfall, they experienced fewer losses than those in many other sheep districts. In subsequent years, especially in the early 1950s period of high commodity prices, high subsidies and the wool bonanza, many large sheep farms were more systematically subdivided into fenced paddocks. Extension officers provided farm-planning advice and veld types were identified as the basis for the division of grazing camps.

One factor inhibiting the multiplication of camps was the difficulty of water provision, since not all had suitable borehole and dam sites. From the 1970s, the use of cheaper plastic piping to distribute water around farms was of major importance in resolving this difficulty, facilitating further fencing. Interviews on sheep farms, especially those which have remained in the same family over several generations, suggest that investment in internal fencing has been a continuous process over many decades. Fences represent a very considerable element in the value of farms which are seldom below 1000 ha, and usually much bigger in the drier districts. Farmers comment that given the low price of land, the cost of fencing from scratch would now equal the price of land. There has been and remains intense debate about the most appropriate forms of grazing, size of camps and frequency of rotation. But it seems likely that rotation systems have been one factor in increasing yields, diminishing selective grazing and reducing the effects of tramping by keeping animals in one place.

Alongside the rotation systems have been developments in pasture improvement and in the use of feed supplements. Some fodder was planted for sheep in the nineteenth century, particularly prickly pear and *Agave americano* from central America. Both were used, chopped, as drought foods. Spineless versions of prickly pear were bred in the early decades of this century, while lucerne spread during the ostrich boom of the late nineteenth and early twentieth century, and remains widely used – for example in raising Angoras – wherever sufficient water can be found. In some areas, lambing is carried out on small areas of irrigated green fodder such as oats, which are also used as winter feed. During the 1930s, in both the USA and South Africa, experiments with planted pastures were carried out, using the best indigenous grasses. Their development has been slow but subsidy programmes in the 1980s have greatly hastened their spread. Various other factors have probably also contributed to veld stabilization. The gradual demise of both white and black tenancy enabled landowners to assert control over the whole area of their farms and institute more systematic farm planning, fencing and grazing regimes. An increase in the size of farms, which roughly doubled in area between the early 1950s and mid-1980s, was certainly associated in the minds of agricultural officials with a decrease in pressure on the veld. Small farmers, desperate to make ends meet, often indebted and with a limited capacity to invest, were seen as particularly prone to overstocking.

While there is no doubt that stock numbers have declined, there is scope for debate as to the causes of this decline and hence its implications for the state of the veld. As noted above, the Drought Commissioners (Union of South Africa, 1923) argued that both the sheep and human populations (they referred to whites only) of some of the key old sheep districts were declining because pasturage and soil were exhausted. The Commissioners were probably not entirely correct in this Malthusian interpretation of the fall in the white population; it was more likely due to the extrusion of white tenants. In many sheep-farming districts, the human population as a whole actually increased in the first half of this century.

A similar, if more sophisticated, argument concerning veld carrying capacity in sheep-farming areas has recently been reiterated by Dean and MacDonald (1994). Using district-level data from the start of the Agricultural Census in 1919, they show stocking rates to have declined overall, especially in the drier districts. By contrast, some grassland districts on the peripheries of the Karoo have experienced growing stock numbers. They cite the negative correlation between increased water provision and numbers as evidence that veld quality is the major constraint. They also note the tendency for farmers to switch from wool-bearing merino types, to the hardier Dorper mutton sheep as evidence of overall decline in pastures. Rejecting various other causes of lower stocking rates, Dean and MacDonald conclude that 'the current livestock stocking rate in the semi-arid and arid rangelands of the Cape Province is unrelated to market forces or state policy but is determined by utilizable primary productivity of rangelands' (1994: 281).

The implication of their argument is that falling stock numbers and reduced stocking rates, which I have cited as a possible factor in explaining improvements in or stabilization of veld quality, reflect an overall decline. Their findings sit uneasily with the botanical evidence of Hoffmann and Cowling. Roux's response to Hoffmann and Cowling is also sceptical. He believes that the thickening of vegetation which has occurred in some locations tends to be of less palatable species so that there is little evidence of increased carrying capacity.[10]

Dean and MacDonald's conclusions, however, are by no means the last word. A number of points require far more systematic historical research. On the one hand, it would have to be shown that the massive investments in fencing and fodder have been at best neutral in their effects and at worst pointless. On the other hand, both state policies and farmers' strategies would have to be discounted as significant reasons for reduction in stock numbers.

It is unlikely that this second point could be sustained in a syste-matic historical analysis.[11] Controlled marketing, price stabilization and subsidies on wool over many decades have almost certainly been a factor in containing sheep numbers. There is no doubt that the huge increase between 1928 and 1931 was largely a response to the collapse in wool prices during the depression, when farmers tried to produce more in order to maintain their incomes and pay debts. Since then, sheep numbers have generally diminished rather than increased when wool prices have fallen. Subsidised destocking programmes, particularly in the late 1960s, have also made some impact. While the fall in stock numbers in the mid-1960s was partly due to drought, the fall at the end of the decade was more likely a reflection of successful government reduction programmes.

The switch to Dorpers and mixed wool/mutton breeds such as Dohne merinos is a complex phenomenon which – in the last few years – has certainly been linked to low wool and high mutton prices rather than simply to veld conditions. Farmers' choices are also influenced by

[10] Personal communication, Grootfontein School of Agriculture, July 1994.

[11] The argument that follows is based on about 25 interviews with sheep farmers and agricultural officers in the Eastern Cape and southern OFS during 1993 and 1994. While these are not considered 'representative' of the whole sheep-farming community, the responses do reveal some general patterns.

factors such as labour costs. In some districts where sufficient water is available, there has been a switch from woolled sheep to beef cattle for labour-related reasons (Beinart, 1994). Fluctuations in the numbers of small stock kept largely for mutton earlier in this century are also unlikely simply to have reflected deteriorating pastures. For example, the percentage of non-woolled sheep (including Black Persians) declined in the first few decades of this century at a time when pastures almost certainly were degrading.

Last, it seems essential that the debate on reasons for the decline in numbers and on the state of the veld also take into account the views of farmers and farmworkers. Not only do they have direct local-level knowledge and experience, but they have also been subject to over a century of debate on degradation, and nearly half a century of government propaganda following the 1946 Soil Conservation Act. Dean and MacDonald are aware of the possibility that government debates have influenced local practice, but do not give it great weight. On the basis of a limited number of interviews in the Eastern Cape and OFS, I would argue that many of the big sheep farmers have inherited their farms and come from wealthier and more educated backgrounds than earlier generations and at least some sheep farmers are aware of environmental issues. They regard with some dismay the stocking practices of their fathers and grandfathers, and are deliberately cautious about stocking levels – a caution which they see as having paid off in the serious droughts of the early 1990s. Most interviewees are of the opinion that the veld has stabilized or improved in many, though not all, districts in their memory.

This interview material is as yet insufficient for hard conclusions to be drawn, and clearly caution is required in dealing with the perceptions of farmers and agricultural officers. Aside from the difficulties in evaluating oral evidence about the general condition of veld in a district, some farmers are clearly aware that political transformations in South Africa may place their rights to the land under question. Arguing that they are now good stewards is part of asserting their rights, as well as their more general belief in private property and the technical ideas that govern their pattern of land use. There are some countervailing pointers in the oral evidence, such as degradation resulting from possible decline in underground water in some districts. At the very least, however, interviews suggest that the views and practices of individual farmers matter and that pasture history must take into account developments on particular farms rather than just generalized arguments about districts and regions. On a number of occasions, I have been shown neighbouring farms where the condition of the veld, separated only by a fence, was significantly different. There are certainly devastated farms, and even areas, but there is also evidence of improvement.

Conclusions

The arguments concerning stock numbers and degradation in South Africa clearly require more extensive debate and research. Botanical work is already rich and is accumulating rapidly (Hoffman, 1993; Dean and MacDonald, 1994; Milton and Hoffmann, 1994; Hoffmann *et al.*, 1995). By contrast, environmental history research which systematically builds in perspectives from political economy as well as ecology has hardly begun. The value of such an approach has been demonstrated elsewhere; for instance in analysis of the Dust Bowl, the transformation of the western prairies, or the regrowth of forest on the Eastern Seaboard in the USA (Worster, 1979; 1985; 1990b). This chapter has tentatively suggested the alternatives to received wisdom which begin to emerge when such a perspective is applied to similar issues in South Africa.

Botanists such as Tidmarsh, although working with a concept of 'climax' vegetation, were aware of the potential resilience of the veld. Ecologists are now increasingly uneasy about cyclical, equilibrium models of vegetation change; it is possible that the Karoo and grasslands on its margins have undergone shifts in vegetation before intensive human exploitation (Milton and Hoffmann, 1994). Once the view from history and political economy is included, it becomes even less helpful to measure environmental change in terms of movement away from a pristine environment and call the result degradation. Human survival requires environmental disturbance. There is no possibility of 'restoration' short of the abandonment of stock-farming completely and – if logical consistency is to be maintained – most other farming in most other areas. Even then, non-equilibrium vegetation changes mean that the species composition of the veld may not be restorable. The evidence from the drier parts of South Africa, as well as many other parts of the country, while by no means clear, suggests that change has not simply taken the form of a linear trend of degradation. While decay is quite possible, a concept of transformation or transition is often more useful.

Some of the well-established discourse about long-term environmental degradation in South Africa, which is enormously persuasive in its general outlines, has been absorbed, perhaps uncritically, by the new radical environmentalist and green lobbies in the country. Much of this analysis, especially with reference to the first half of this century, may be accurate. But it is also an attractive instrument with which to condemn the past.

In planning for the future, it is arguably important to begin to disentangle apartheid, and the white farmers who largely supported it, from the environmental condition of the farms. Apartheid has certainly contributed to environmental degradation in homeland areas (although even here the picture in some zones may be less bleak than it is sometimes painted). But the cosy relationship between organized agriculture and the apartheid state over many decades may in fact have facilitated conservationist policies on the white rangelands. Cowling's vision of a recommunalized Karoo with nomads, roving flocks and game, may be an attractive social scenario, but could also reintroduce a degree of unpredictability in an area where there has been some evidence of constrained exploitation under systems of private tenure. Further undermining of commercialized sheep farming (whether by white or black farmers) may also have adverse social consequences, not only for landowners, but also for the small towns and farm-workers of the region. And while reintroduction of game is an important option, its social and ecological effects are highly uncertain and depend on the way it is implemented.

4

Desertification | Narratives, Winners & Losers

JEREMY SWIFT

Introduction

The story of 'desertification' can be read in several ways. In this chapter, I document the history of the concept, and argue that it has less to do with science than with the competing claims of different political and bureaucratic constituencies. 'Desertification' is perhaps the best example of a set of ideas about the environment that emerge in a situation of scientific uncertainty and then prove persistent in the face of gradually accumulating evidence that they are not well-founded. This 'stickiness' of ideas has important implications for policy.

Although the word 'desertification' has been widely used since the 1950s, an ill-defined, changing set of definitions has been used as shorthand for the concept, and hence the range of ecological and other processes included within its bounds has constantly shifted.[1] At the risk of adding to the confusion, in the rest of this paper I use the term 'desertification', without inverted commas, to mean the generally received wisdom that dryland environments are being rapidly degraded (reduced in long-term economic productivity) by a varying mix of natural and human factors. Because this is largely a narrative construct, I refer in places to this as the received desertification narrative, in Emery Roe's sense (Roe, 1991). As will be clear, I am sceptical about the narrative; the aim of this chapter however is not to show whether or not certain processes are happening (although I summarize recent evidence about this), but why a particular set of poorly researched ideas has been so influential, and why the received desertification narrative has persisted as a beacon to policy in the face of so much counter-evidence about its key premises. In doing so I ask the question as to who are the winners and losers from the received narrative, and whether a better counter-narrative is available.

---

[1] Glantz and Orlovsky (1983) review more than a hundred definitions of desertification; the number has certainly increased since.

A short history of desertification

Stebbing and the West African colonies

The process now called desertification (the word itself was not yet used) first came to prominence in West Africa in the 1920s and 1930s, as a result of concerns by French colonial administrators, foresters and social scientists about the apparent progressive drying out of large areas of their Sahelian colonies, and a presumed extension of the Sahara.[2] In the 1930s these worries spread to the English-speaking colonies. The work of E.P. Stebbing was especially influential. Stebbing, a forester from the Indian Forest Service, made a tour in 1934 through British and French colonies in West Africa and up across the Sahara, following the transition from mixed deciduous forest near the coast, through various types of savanna, out into the desert itself.

Stebbing wrote prolifically about this trip, producing a book and several long articles, which summarized French studies and reported detailed observations of his own.[3] He was clear in his own mind about what he had seen:

> The present-day results of investigations would appear to prove that the Sahara is far from stationary on its southern frontiers; that blown sand and desiccation are increasing in the colonies lying in juxtaposition to the desert, and that the present method of agricultural livelihood of the population living in these regions, with their unchecked action of firing the countryside annually, and methods of pasturage – all tend to assist sand penetration, drying up of water supplies, and desiccation. (Stebbing, 1937b: 31)

The following year, he spelled out in more detail the processes he believed to be at work.

> In West Africa the process [the advancing Sahara] owes its commencement to the system of farming the bush or degraded type of forest which covers much of the countryside, this system being a form of shifting cultivation. With an increasing population the same areas are refarmed at shorter intervals, with a consequent more rapid deterioration of the soil constituents, until a stage is reached when the soil is no longer sufficiently productive for agriculture.
>
> It may then be made over for stock-raising. The grazing and browsing, accompanied by the universal practice of annually firing the countryside, reduces the quality, height and density and therefore the resisting power to sand penetration of the now much-degraded forest or bush, and the soil becomes covered with a sandy top, gradually increasing in depth. Once this stage has developed the rapidity in the degradation of the bush increases until it no longer affords sustenance to cattle; then the sheep disappear; and under the

---

[2] The debate from this period is described in several sources, and is well summarized by Mortimore (1989: 12–15).

[3] Stebbing 1935, 1937a, 1937b, 1938, and several others.

final exploitation by goat herds to feed their flocks the savannah suc-
cumbs and the desert has encroached and extended its boundaries.
(1938: 12–13)

The prime factors in this process, he wrote, are indigenous forms of
land use, made more damaging by population growth resulting from the
ending of local warfare and improvements in health, resulting in a
reduced area available for cultivation, and a large increase in animal
numbers (1937b: 33). He also postulated that there was a feedback from
damaging land use to reduced or more intermittent rainfall, although he
did not specify how this mechanism might operate.

Stebbing gives one of the earliest estimates of the rate of southward
advance of the desert, quoting a retired French colonial administrator
who believed that in Mali and Niger the Sahara had advanced at an
average of one kilometre a year during the previous three centuries
(1937a: 24); he also reported that desert advance was accelerating (1938:
24).

Stebbing's work was influential, and the main features of his analysis
are found in much writing about desertification today. They include:
(i) the desert has been expanding southwards, at a measurable rate, for
several centuries; this rate has recently speeded up; increasingly mobile
sand dunes and dust storms are a major part of the problem;
(ii) 'desiccation' is mainly to blame: desiccation means the drying up of
surface water, a lowering of the water table, and a decrease in rainfall;
(iii) desiccation results from human misuse; especially shifting
cultivation, the extension of cultivation, greatly increased nomadic
grazing pressure and tree lopping by herders, and bush fires; these act
directly on surface and soil water availability and feed back negatively
to rainfall; (iv) human misuse is in turn the result of human population
increase, resulting from an ending of local warfare and an improvement
in health; (v) climate change and rainfall variability are specifically
rejected as causes; rainfall has, as a result of erosion and vegetation
degradation, become 'capriciously intermittent' or more variable, but
this is a result, not a cause; and (vi) West Africa is not the only
vulnerable place: eastern and southern Africa, the US dust bowl area
and India, are subject to the same processes.

Stebbing argued that to face such a menace, an international response
was needed, involving cooperation between the British and French
colonial authorities. The answer lay in forest reserves:

In order to stem the southward progress of the sand a belt of high
forest should be recreated along the international border by reserving,
on a 15–mile depth, areas as continuous as possible *of the existing
degraded dry mixed deciduous forest*. That in this belt no firing
should be allowed, and that cultivation and the pasturage of herds
and flocks should be restricted, and eliminated altogether where
possible (Stebbing, 1937a: 29, emphasis in original).

He proposed a second, parallel, forest belt further south through central Nigeria, Dahomey, Togo, Ghana and Ivory Coast.

Elsewhere Stebbing argued that an extension of central and local government authority over natural resource use would be necessary:

> . . . regulation of the farming, prohibition of firing the countryside except by permission of the local Authority, and conservation and protection of all important forest areas [are required]. Only the Central Authority can introduce such measures with the recognition that they are for the good of the community as a whole. (Stebbing, 1938: 40)

The Anglo-French Forestry Commission

Stebbing demanded that a small international group of scientists should investigate the problem urgently, and this was accepted by the authorities. The Anglo-French Forestry Commission was set up, and did extensive fieldwork in the cold season 1936–7 both sides of the Niger–Nigeria border between Niamey and Lake Chad.[4]

The Commission's trip followed an exceptionally wet 1936 rainy season, which gave them a different perspective from Stebbing and earlier observers. It was apparent to them that the supposed lowering of the water table and general drying out identified by Stebbing was not taking place. They pointed to the variability of rainfall:

> It seems that dry and wet periods, of short and variable duration, follow each other. They do not demonstrate any tendency towards a permanent change in climate. The vegetation follows this rhythm, with regeneration taking place readily in the wetter years, but with greater difficulty in dry years. (Anglo-French Commission, 1973: 10)

The Commission systematically refuted most of Stebbing's assertions. No large-scale sand movements were recorded; with four very local exceptions, ancient dunes were found to be anchored by grass and woody vegetation, and no agricultural land was threatened by sand. Despite some areas of dead trees, abundant forest regeneration was seen to be taking place, including some on dune soils. It was impossible to generalize about the state of the woody vegetation: in some places it was retreating, in others advancing. In general, there had been an extension of shifting cultivation to the detriment of forest vegetation on a large scale; the Commission was of the opinion that intensified, permanent agriculture, of the sort observed around Kano, was more satisfactory and should be pursued as a goal elsewhere.

The Commission's recommendations were considerably more modest than Stebbing's: the plantation of shelter belts and woody hedges, and

---

[4] 'Report of the Anglo-French Forestry Commission 1936–37', Nigeria, Sessional Paper No. 37 of 1937, Lagos. The French version of the report was published in Anglo-French Commission (1973). I have worked from this French version in the following discussion, and quotations in the text are my translation from the French version.

the protection of trees, especially *Acacia albida*, in fields; a practice already followed by farmers in many areas.

Several members of the Commission published scientific papers based on their observations, and the controversy was summed up in 1940 in an important paper by Dudley Stamp, a geographer (Stamp, 1940; see also Jones, 1938). Stamp supported the Commission's findings, pointed to several shortcomings of Stebbing's analysis, including especially his assumption that savanna vegetation was degraded forest, lamented the lack of interdisciplinary analysis in what was essentially an ecological matter, and concluded that there was little evidence for desiccation as an autonomous mechanism, but that such a problem as existed was mainly one of soil erosion.

> There now seems to be little doubt that the problem before West Africa is not the special one of Saharan encroachment but the universal one of man-induced soil erosion, which necessitates remedial measures comparable with those being adopted in other parts of the world but with special modifications in view of the local agricultural system of bush fallowing and burning. (Stamp, 1940: 300)

The discussion continued in these terms through the 1940s and 1950s. However Stamp's warning not to confuse the observed change in vegetation from forest to grassland along a transect from wetter to drier areas with a process of progressive degradation at the drier end was largely ignored. The term desertification was coined in 1949, by A. Aubréville, a French forester who had been a member of the Anglo-French Commission, to describe what he saw as an almost entirely man-made process of destructive land use across the whole forest and savanna zones of West Africa. His views seem to have changed after the Commission's report was presented, and the tone of his writing, and that of other foresters, became increasingly critical of the land use practices of farmers and herders.

Deserts on the move again

After a period of quiet in the 1950s and 1960s, which were generally wet years in the Sahel, the concept of desertification was revived in the early 1970s, triggered by the major drought across much of the Sahelian and Sudanic ecological zones, which started in the late 1960s and peaked in 1973 in West Africa and 1974 in the Horn. One of the earliest reactions was a review by the Office of Science and Technology of the US Agency for International Development (USAID) which stated, without sources, that

> . . . measurable encroachment in northern areas of [the Sahelian] countries has occurred and is continuing. While specific areal data are lacking, a rough estimate of magnitude of encroachment south of the Sahara is that about 150,000 square miles of arable land (i.e.

suitable for agriculture or intensive grazing) has been forfeited to the desert in the past 50 years . . . several studies of the Sahara have concluded that there has been a net advance in some places, along a 2,200 mile southern front, of as much as 30 miles a year. (USAID, 1972: 2–4)

Of the research carried out in the 1970s, the most important was in north-western Sudan. Two investigations were especially influential. In 1975, a short ecological reconnaissance was commissioned by the Sudanese government, UNESCO and the UN Environment Programme, and carried out by Hugh Lamprey, an experienced ecologist and wildlife biologist. About three weeks work was done in late 1975, combining ground survey and low level aerial reconnaissance, although the most influential report was that summarizing Lamprey's air survey (Lamprey, 1975). This report remained unpublished – indeed was treated as confidential – for at least a decade after it was written, but its conclusions were widely cited.

They were dramatic. Lamprey compared his observations on ecological zonation and the location of the desert edge with those of a botanical survey carried out in 1958, and concluded that: (i) in north Kordofan and Darfur, ecological boundaries had shifted south; the desert boundary had moved south by about 90–100 km in the 17 years between the two surveys; (ii) the ephemeral winter vegetation known as *gizu*, characteristic of the southern desert edge, had shifted some 80 km south of its last reported occurrence in 1964; (iii) sand encroachment was taking place as desert sand was blown south on a broad front; in some areas sand had moved rapidly ahead of the desert and was accumulating on formerly consolidated soils, killing vegetation, and threatening farm land and villages; (iv) there had been extensive mortality of gum arabic (*Acacia senegal*) trees along the 14th parallel; (v) there was a progressive abandonment of agriculture in the North, particularly where sand encroachment had taken place; (vi) in the northern Nile valley, the alluvial agricultural strip along the river was being reduced by drifting sand, which was obliterating farm land and villages.

A geographer, Fouad N. Ibrahim, worked on the same topic in northern Darfur, Sudan, in 1976–7, with observations continuing until 1982 (Ibrahim, 1984). The opening paragraph of his book, based on his doctoral thesis at the University of Bayreuth in Germany, sets the tone:

The often reported advance of the desert on all continents lowers the yield capacity of the earth and deprives the populations in the affected areas of their nutritional basis. Worldwide, about five million hectares of land are lost each year to desertification, another 30 million square kilometres, i.e. a fifth of the land area of our planet, are threatened by the spread of the deserts or have already been partially affected. The presently observed advance is called desertification. It is a process which slowly destroys the regenerative capacity of the vulnerable ecosystems in arid and semi-arid regions with land use methods which are not adapted to natural conditions . . . In the past fifty years, the Sahara alone has taken over 650,000 square kilometres through desertification of the Sahel . . . man is the real cause of this desertification. (Ibrahim, 1984: 17)

The work of Lamprey and Ibrahim, and others reaching similar conclusions, played a key role in forming an international consensus in the late 1970s and early 1980s about desertification. The problem was seen to be mainly the result of human populations higher than the capacity of the land to support them, and inappropriate forms of land use; depending on the point of view of the commentator, these could be arable cultivation in areas too dry for sustained productivity, extensive pastoralism with herds above the carrying capacity of the pastures, or indiscriminate cutting of firewood. The goat was a major villain. Periodic drought was thought to exacerbate these imbalances, but not to be a main cause. The main physical manifestations of desertification were declining biological productivity of the land, deforestation, and an increase in mobile sand dunes. Although most commentators argued that the problem was not so much the desert moving outwards as the creation of islands of degradation in the savanna which spread and eventually linked up, the images and metaphors used in most reports remained those of the desert on the move. Lamprey's estimates for northern Sudan (the desert edge had moved south 90–100 km in 17 years) was interpreted as an average annual advance of 6 km, and this figure became the basis for increasingly unlikely rates of desert expansion along the whole southern Saharan edge.[5] This consensus led to concerted international action for the first time since the Anglo-French Commission forty years earlier.

UN Conference on Desertification
International concern about what was happening in the drylands was mobilized on a large scale during the 1970s, culminating in the UN Conference on Desertification (UNCOD), held in Nairobi in 1977.

UNCOD went about its task in some detail. Four extensive reviews of the existing state of scientific knowledge were commissioned (on the relationship of desertification to climate, ecological processes, society and technology), as well as a comprehensive set of desertification maps and national or local case studies of desertification.[6]

---

[5] Such as the World Bank President's claim in 1987 that in Mali the Sahara had moved 350 km south in the previous 20 years (Conable, 1987: 6).

[6] The main case studies were commissioned by the UNCOD secretariat; other case studies and national reports were volunteered by countries. The latter had some surprising instances of contemporary desertification – including Cyprus, Portugal and Zaire – as well as some countries where presumably the problem was not new, such as Jordan, Kuwait, Libya and the United Arab Emirates.

Many of the scientists involved in UNCOD were uncertain about the causes and extent of desertification. But the final report of UNCOD ignored their pleas for caution. It argued that:

> . . . the problems of desertification are larger, more widely shared, and require greater and longer term action than expected. The simplistic fears of a few years ago are now replaced by a well-founded sense of danger . . . desertification is not a problem that faces just a few countries. (UNCOD, 1977a: 1)

The Conference estimated that more than a third of the earth's surface was natural desert or semi-desert, that an additional 10 per cent was man-made desert, and that a further 19 per cent was threatened with desertification, with this threatened area distributed in over 100 countries (UNCOD, 1977a: 2). The main threats came from increased intensity of land use, especially removal of the natural vegetation cover by farming, overgrazing and incorrect irrigation, exacerbated by drought. The advance of the desert itself was not considered to be the main problem, so much as gradually enlarging patches of degradation which eventually join up; however, an apocalyptic vision was presented of an inexorable and almost contagious process:

> Where pressure of land use persists through drought, these same ecosystems are shown to be fragile, and processes can be set in motion whereby desertification becomes self-accelerating. This can occur where sand dunes are stripped of vegetation . . . and drifting sand destroys more vegetation and mobilises extending surfaces, and dunes slowly advance and engulf less damaged sites . . . Because self-acceleration can occur through a variety of circumstances, desertification will often advance inexorably unless preventive measures are undertaken. As it advances, it becomes ever more difficult and more expensive to treat, with the costs of reclamation continually rising until the stark equilibrium of extreme desert is reached and the land has for all practical purposes passed beyond hope of rehabilitation. (UNCOD, 1977a: 5)

UNCOD drafted a Plan of Action to Combat Desertification, to be implemented by the year 2000, with the overall goal of preventing and arresting 'the advance of desertification' (UNCOD, 1977a: 7). The plan contained 28 detailed recommendations, and also proposed five large-scale transnational development projects to manage the major regional aquifers of the Horn of Africa and the Arabian peninsula, 'green belts' from the Atlantic to the Red Sea and Indian Ocean across the northern and southern Saharan edges, a regional livestock project along the whole Sudano-Sahelian belt, and regional desertification monitoring in South West Asia and South America. The Sahelian green belt, variously conceived as 50–100 km wide or the whole depth of the Sahelian zone, would be managed more conservatively and increased forestry activities undertaken; it was explicitly described as a revival of Stebbing's proposed forest belt. The northern Africa green belt was to be narrower than the Sahelian one, with more attention to tree planting, but would also be the width of Africa (UNCOD, 1977b; 1977c; 1977d).

The UN Environment Programme was given a mandate to take forward the Desertification Conference's work, and reviewed progress in implementing the desertification action plan in 1984. To assist this process, UNEP commissioned several surveys. The centrepiece was a printed 29–page questionnaire, sent to 91 countries thought to be affected, asking in great detail for information on indicators of human population, changes in land use, and crop livestock production. Unsurprisingly, many countries felt they did not have adequate statistics, but were encouraged to make estimates, which were processed by UNEP in Nairobi.[7] Those analysing the data had doubts about their validity (see for example Berry, 1983: 4–5), and the group of outside advisers asked to provide an overview had strong reservations about the process. But these doubts did not survive the editing process in the preparation of the final report of the Executive Director for the UNEP Governing Council (UNEP, 1984). This document summarized the results in sombre terms:

> The scale and urgency of the problem of desertification as presented to the Desertification Conference and addressed by the Plan of Action have been confirmed. Desertification threatens 35 per cent of the earth's land surface and 20 per cent of its population; 75 per cent of the threatened area and 60 per cent of the threatened population are already affected through deterioration of the environment and living conditions, and between a quarter and a half of the affected population severely so. In the years since the Conference, the land irretrievably lost through various forms of desertification or destroyed to desert-like conditions has continued at 6 million ha annually as reported in 1977; and the land reduced to zero or even negative net economic productivity is showing an increase over the 1980 estimates, at 21 million ha annually. (UNEP, 1984: 17)

These figures were destined to become a key part of the desertification narrative. For example, they were used, without attribution of source, by the Brundtland Commission as a key example of the severity of the global environmental challenge (WCED, 1987: 2). Many later uses, if attributed at all to a source, use Brundtland. This UNEP view became the received wisdom on desertification, scarcely challenged in public policy-making or popular reporting.

But the received view had never been accepted by many dryland scientists. The data and analysis outlined above contain serious flaws

---

[7] In at least some cases, the results were surprising. I understand that in Bangladesh the form was filled in by a new remote sensing institute, without ground-truthing; they compared 1977 images taken during the rice growing season with 1983 images of rice fields prepared but not yet planted, and concluded that there had been a large increase in bare ground (personal communication from someone present at this exercise).

and dubious generalizations. These have been subject to considerable analysis.[8]

Alternative views

Some of the confusion arises because the term desertification has been used to cover three related but distinct phenomena: drought, dessication, and dryland degradation (Warren and Khogali, 1992: 6–7). These distinctions are partly practical, partly scientific:

- Drought, defined as two or more years of rainfall well below average, normally depresses primary production substantially, especially in drier areas where there is a close relationship between available soil moisture and plant growth. With better rains, the vegetation usually recovers quickly to its pre-drought level, without any long-term damage to productivity. Drought pulses of this sort are now seen as one of the key forces driving dynamic ecological systems; droughts lead to great variability in ecosystem processes and productivity, not to a secular decline in productivity. In Sudan, large inter-annual swings in biomass production, as measured by satellite imagery, can be explained by variations in rainfall (Helldén, 1991).
- Desiccation is a process of more general drying out, resulting from extended drought, on the order of decades or more. Desiccation may have more substantial ecological effects, especially where trees die due to lack of soil moisture, but even this sort of change is likely to correct itself over a longer time horizon.
- Dryland degradation is a persistent decrease in the productivity of vegetation and soils, brought about largely by inappropriate land use leading to physical changes in soil or vegetation structure, irrespective of levels of rainfall or soil moisture. Because of the masking effects of drought and dessication, it is often very difficult to detect where degradation has taken place; for example, it is now known that much of the degradation assumed to have taken place in dry grasslands as a result of overgrazing is a misreading of the situation, resulting more from the immediate effects of drought, and reversed rapidly in conditions of more normal rainfall. Some forms of dryland degradation, however, especially accelerated soil erosion by water or wind, may be effectively irreversible.

The advice of many dryland researchers for great care in distinguishing between these processes, and in interpreting survey data on ecological status, especially where there were few long time series data, was

---

[8] See for example Nelson (1988), Rhodes (1991), Warren and Khogali (1992), Mortimore (1989), and especially the extensive work of the group at Lund University, Sweden, led by Ulf Helldén, conveniently summarized in Helldén (1991).

generally ignored in the drafting of the ambitious UN plans, as was the general perception of farmers and herders that highly variable rainfall and a long drought at this time were the main causes of the observed low biomass productivity. But when researchers argued that the science was not well understood, they were accused of standing in the way of aggressive solutions to urgent and well-known problems.

Some of the scientists who prepared background documents for UNCOD were already sceptical about what was clearly destined to be the received narrative at UNCOD, and argued for great caution in interpreting the small amount of useful information available. The ecology background paper, the prime source of scientific information on the processes the conference was debating, contained a proper note of caution:

> The evidence for desertification is diffuse and almost impossible to quantify. There can be little reasonable doubt that many environments have suffered serious damage, and this is mostly by cultural practices, but the persistence of the effects is much more debatable. In particular, the losses that followed recent droughts in the Sahel, and elsewhere, cannot be classified as desertification until clear evidence emerges that the yield of useful crops in the ensuing good years has been depressed beneath the preceding wet period, and that the losses are due to environmental causes. (Warren and Maizels, 1977: 19)

As we saw earlier, the final report of UNCOD ignored this plea.

Part of the problem arose from the fact that during the 1970s and 1980s there was very little good primary field research on the processes of dryland degradation.[9] A small number of works were repeatedly cited by analysts and development agency people; but these, especially the key works by Lamprey and Ibrahim described earlier, have important weaknesses. The most important is that dry years (especially in the 1970s and early 1980s) were often compared with wet years (especially in the 1960s); this was then interpreted as a secular decline in productivity, rather than as a variation in the response of the natural vegetation or crops to soil moisture availability. Ibrahim also used declining millet yields per hectare as proof of declining land productivity; in fact his yield figures clearly follow the pattern in rainfall in the millet production areas, with an additional depressing effect on yields resulting from the extension of millet cultivation into drier areas which he himself described.

---

[9] In 1987, I wrote to around 50 researchers active in arid lands asking for their top ten list of scientific papers on desertification; most replied, but few had more than one or two papers to propose, and the total of reputable works from the exercise was no more than two or three; the Desertification Branch at UNEP suggested three papers, two of them unobtainable 'grey' literature, the third concerned with desertification in Algeria during the Neolithic.

Despite these shortcomings, an alternative picture was already emerging in the 1980s. The most comprehensive research, combining remote sensing, extensive field observations, national economic statistics and modelling, was carried out over several years in Sudan by geographers from Lund University in Sweden (Helldén, 1991). The conclusions of this work run directly counter to the received narrative of desertification.

> None of these studies verified the creation of long lasting desert-like conditions in the Sudan during the 1962–1984 period, which corresponded to the magnitude described by many authors. There was no trend in the creation or possible growth of desert patches around 103 examined villages and water holes over the period 1961–1983. No major shifts in the northern cultivation limit were identified, no major sand-dune transformations or Sahara desert encroachment trend was identified. No major changes in vegetation cover and crop productivity was identified, which could not be explained by varying rainfall characteristics . . . There was a severe drought impact on crop yield during the Sahelian drought 1965–1974 in the Sudan followed by significant recovery as soon as the rains returned. (Helldén, 1991: 379)

The same author concluded that, although there was a shortage of reliable data, the hypothesis of a secular, man-made trend towards desert-like conditions could not be confirmed in other dryland areas of Africa either (Helldén, 1991: 383).

In West Africa, Mortimore (1989) investigated in detail the Manga Grasslands, a prime candidate for degradation since it is an ancient fixed dune system with some moving dunes on the Niger–Nigeria border, densely inhabited by a mix of farmers and herders. Mortimore was able to compare aerial photos from 1950 and 1969, spanning a period of economic expansion, and growth in population, agricultural production and settlement. The air-photo analysis was accompanied by detailed ground investigation. Although stressing caution in the interpretation, Mortimore was able to come to clear conclusions about changes in land use and productivity: woodland had deteriorated, with cutting exceeding regeneration; grassland had undergone a change in floristic composition, but no necessary reduction in forage availability; no conclusive evidence of soil degradation other than some movement of surface materials; increased size and frequency of rangeland dunes, due at least in large part to declining rainfall. The boundaries of the Manga Grasslands in 1969 were in exactly the position they had been in 1937, when they were described by the Anglo-French Forestry Commission.

Mortimore concluded (1989: 185–6) that

> The stability of the borders of the Grasslands, their largely treeless flora, and the known existence of moving dunes since 1937 shows that the basic characteristics of the area have some continuity in

time. Most of the evidence for ecological degradation derives from the last two decades, corresponding to a decline in the rainfall.

New realism

These data, the reservations of most dryland scientists about the received narrative of desertification, and a changed leadership at UNEP, have led in the last five years to a new realism about desertification, although two strands of thought – the one based on the old alarmism, the other linking dryland degradation much more closely to soil erosion and productivity variations due to the rainfall variations – now coexist uneasily. Desertification was debated at the Earth Summit in Rio, and chapter 12 of *Agenda 21* is devoted to it (UN, 1992). The analysis still uses some of the old figures about different categories of land loss, but gives climatic variation as a cause on an equal footing with human activities. Dryland populations and especially poverty are now central to the analysis and to the solutions. The measures proposed are closer to the ground, more modest, and admit considerable uncertainty. Improved information and understanding, better soil conservation and afforestation, a poverty focus and a search for alternative livelihoods, as well as drought preparedness, are among the recommended solutions, and there is a stress on participation and environmental education. The need for changes in national policy in order to provide a better enabling framework and incentives for the activities of herders and farmers is fully recognized.

Following *Agenda 21*, an international Convention to Combat Desertification was signed in 1994, with a stress on improving the livelihoods of dryland inhabitants. The presence of experienced researchers in the preparations for these negotiations, and the changing wider climate of opinion about desertification, has meant that the concepts enshrined in the new draft convention are far removed from the simplistic formulations of the received narrative.

There are few scientists or international administrators now who would defend the received narrative of desertification, although it lingers on in many government departments in dryland areas and some development agencies, and is often raised as a critical issue in project formulation in dry areas. A simple idea, adorned with powerful slogans, proves remarkably hard to change, even when shown to be patently inaccurate.

Desertification as development narrative

Two general points should be made about desertification before entering a wider discussion. First, an interest in understanding the received narrative about desertification does not imply a belief that there is no problem of land degradation in the drylands. On the contrary, it is arguable that the received narrative about desertification has for many years stood in the way of more effective approaches to soil erosion and land degradation generally, by focusing attention on poorly defined problems and misguided solutions. There are potentially serious environmental problems in the drylands, especially soil erosion. The desertification debate has distracted attention from them, and one aim of understanding it better is to return attention to more urgent problems and more realistic solutions.

The second point is about historical relativity. It is certainly no accident that both periods of intense discussion about the encroaching Sahara started in or shortly after periods of drought. In West Africa, French concern about desertification started in the context of low rainfall in several years between 1905 and 1920, especially 1913, a major drought and famine year (Mortimore, 1989: 13). The surge of interest in desertification and the intense UN and other donor activity in the 1970s and 1980s started during perhaps the driest decades of this century. In such conditions, it was easy for civil servants, colonial administrators, foresters and journalists to assume that Africa was drying up. The whole desertification debate shows that it has been hard even for some researchers to distance themselves from the effects of a single year's rainfall. If cyclical variability in rainfall in the drylands now produces a period wetter than the last two decades, many of the obvious physical manifestations of desertification will disappear under an excellent grass cover and productive millet fields. It is important that concern for dryland environmental problems does not disappear with them.

However, a misreading of climatic variability is not enough to explain the tenacious hold of the received desertification narrative on the minds of politicians, civil servants, aid administrators and some scientists. The persistence of this narrative, in the face of considerable scientific scepticism from the start, and accumulating evidence to the contrary, requires some explanation. Clearly the narrative meets a need, and provides a useful discourse for someone. In the remainder of this chapter I argue that the received narrative of desertification provided a convenient point of convergence for the interests of three main constituencies: national governments in Africa, international aid bureaucracies, especially United Nations agencies and some major bilateral donors, and some groups of scientists.

National governments

National governments in Africa in the 1970s were searching for a justification to maintain their preeminent position in rural development, and to rescue an ideology, already failing at that time, of authoritarian intervention in rural land use. 'Desertification' was the crisis scenario they used to claim rights to stewardship over resources previously outside their control (cf. Roe, 1995).

The Stebbing controversy of the 1930s, with its roots in French colonial botany of the previous decades, took place at a time when France, having recently completed the conquest of the Sahelian countries, was setting up and seeking to justify a large and highly centralized natural resource management bureaucracy on the French metropolitan model. The assumption that local herders and farmers had such inefficient systems of land use that they were destroying the land was, at the very least, a convenient one at such a moment. Periods of drought reinforced that perception for outside observers unused to the great inter-annual variability of the Sahelian climate. It is perhaps significant that it was in general soldiers, administrators and foresters (the most military of civil servants in the French system) who argued the desertification case most strongly, and geographers, geologists and other scientists who were most sceptical.

It seems likely that renewed alarmism about desertification in the 1970s owed something to the same factors. The governments of many African countries, recently independent, were recasting their bureaucracies at this time to give them more extensive and centralized powers over the administration of natural resources, and over the inhabitants of the drylands, especially the pastoralists who were seen as subversive of national unity. The droughts of the early 1970s, the worst since 1913, made the landscape look degraded enough for such claims to be plausible to many outsiders. Unwise speeches by senior international figures, about the threat desertification posed to the very existence of some African dryland countries, added to this legitimacy.

Desertification justified increasing control by natural resource bureaucracies, such as the planning, forest and wildlife or national parks services, over land, and those within and without such services who benefited by such arrangements were among the most vocal in support of national anti-desertification plans. The UNCOD Plan of Action made improved land-use planning central to its recommendations, and envisaged a large extension of control by central planners over rural land use. According to the Plan

> . . . a comprehensive land-use plan would assign all sections of the area to particular uses, such as crops, livestock, game ranching, forests, biosphere reserves, recreation . . . where planners determine that a section of land is critically endangered or has become unsuitable for human activities, they should propose a degree of protection, ranging from complete withdrawal to limited uses which promote natural recovery. This is particularly important in areas recently subject to severe degradation under the impact of human activities. (UNCOD, 1977a: 11–12)

There were little more than perfunctory nods in the direction of the logic of existing land uses, and local participation in planning.

Pastoralists came in for special blame. Although the UNCOD Plan of Action performed a delicate balancing act between those (mainly governments) who wished to settle pastoralists, and those who considered mobility an advantage in using extensive and variable range-lands, national plans to combat desertification generally took a simpler view: pastoralists were prime culprits in the 'tragedy of the commons', structurally unable to manage the land conservatively; their goats were especially damaging and they themselves lopped and felled trees indiscriminately; and their irrational attachment to livestock numbers and unwillingness to sell animals quickly led to overgrazing in the fragile marginal dry environments of the desert edge. This 'mainstream view' (Sandford, 1983) underlay most national plans about the rangelands and their extensive pastoral economies in the 1970s and 1980s, and to some extent still does. The desertification narrative provided governments with an internationally accepted excuse to be nasty about and to pastoralists.

In extreme cases, the desertification narrative, with its implicit threat to the survival of whole dryland economies, was used to justify politically authoritarian actions. Senior politicians sometimes found it convenient to invoke desertification in support of their activities. In 1984, President Kountche of Niger used the need 'to fight against the advancing Sahara' as the context in which he cracked down on merchants who stole food aid and to sack 30 traffic police. In 1985 he went further: calling on citizens to step up their fight against the advancing Sahara desert, he shelved plans to liberalize the domestic political system. 'We cannot talk politics on an empty stomach', he announced (Warren and Agnew, 1988).

Aid bureaucracies
The desertification narrative was also a useful justification for calls for increased aid flows. The need for increased funding for anti-desertification measures was a constant theme of governments, especially in Africa, during the 1970s, and when the UNCOD Plan of Action was not implemented, the blame was put on inadequate funding rather than inadequate ideas. In the 1970s, following the Stockholm Conference, environmental arguments began for the first time to make real progress in international negotiation, and Southern governments and international organizations were not slow to seize on them to argue for more aid.

For the aid agencies, desertification seemed in 1977 an ideal theme for their activities, since it was seen to lie largely outside the political arena (unlike poverty, land distribution or birth control, all raised during UNCOD, but sidelined because of their controversial nature), without powerful losers. As in the case of national governments, desertification provided a crisis narrative enabling aid agencies to assert rights as stakeholders in the drylands. The scale and nature of the perceived threat provided a perfect justification for large, technology-driven, international programmes of the sort large multilateral aid agencies thrive on and which were then becoming popular: the UNCOD Plan of Action included five huge international programmes of this sort, which needed an international agency to manage them. They provided an excellent flagship programme for UNEP, a new UN agency struggling to find a niche.[10]

The famines of the early 1970s, which captured enormous attention in the media and in aid agency thinking, also contributed to this. The trouble was that famine had inconvenient political ramifications, and was regarded as a political minefield by donors. There was no major programme, no UN Plan of Action, to combat famine for this reason. Desertification, on the other hand, was seen as related but politically safe, and a lot of the feelings of guilt, and the energy and resources, of donors, were channelled into desertification as a surrogate for doing something about famine.

Scientists

For some scientists also, desertification provided a shot in the arm. In the 1970s, ecologists were just beginning to realize they could be involved in policy, and this seemed an ideal issue to ride, since it promoted the image of ecology as useful and not politically sensitive. The US range managers' view – that most African pastures have been degraded from the desirable climax vegetation by overgrazing – was supported by the desertification argument, and such people played a key role in USAID rural projects at that time. Remote sensing, in its infancy, could perhaps provide the answers, without too much need for difficult and lengthy ground work, and desertification provided a justification for large-scale funding for the development of remote sensing in Africa.[11]

Scientific research, especially social science involving lengthy fieldwork, was always seen as inconvenient by the proponents of desertification in government and aid agencies: too long, too detailed, too likely to come up with complex stories with no clear message. Even after the proper research began to appear, such as the Lund University Sudan data from the mid-1980s onwards, the main international organizations involved in desertification, like UNEP, ignored a counter-narrative they found inconvenient.

Conclusion: a counter-narrative?

To explain the desertification narrative in terms of the convergent interests of governments, aid agencies and scientists is not a form of

[10] In fact, none of these projects were funded, donors doubting UNEP's ability to manage them.

[11] One of the most important stimuli to the development of both remote sensing and the desertification debate in Africa was a satellite image during the 1971 drought in the Sahel of a square of green amid the sand; enquiry showed this was a fenced state ranch at Ikrafane in Niger, destocked early in the drought to provide grass for the President's cattle. On this basis, fenced ranches were widely promoted by the US aid community as the answer to both desertification and famine.

conspiracy theory; it is an attempt to explain policy outcomes which are otherwise baffling. The desertification story is a particularly interesting example of this because the received narrative persisted in the face of rapidly mounting scientific evidence that it was inaccurate, and that the policies it suggested did not deal effectively with dryland degradation. We may explain this by understanding who were the main winners from the received narrative, and what they won: a measure of legitimacy in making decisions over dryland resources. The fact that in general these decisions, when they were not irrelevant, were inefficient, and often quite harmful to sustainable dryland management, is beside the point. The narrative established the right of the winners – national governments, aid bureaucracies and some scientists – to participate in these decisions and to try to impose their view. In terms of institutional logic, these were important gains.

There were also a clear set of losers from this narrative, although they did not pack much of a punch: dryland farmers and herders, whose own control over resources was whittled away by central planning, land tenure reform, ranches and other good ideas from governments, the aid agencies and outside consultants.

It is the fate of development narratives to be replaced by counter-narratives which in their turn are replaced by counter-counter-narratives. In the case of the drylands, there is already the outline of a persuasive counter-narrative, more attuned to the concerns and strengths of the losers, more plausible, more participatory, based on better science, and more likely to result in better land management. This counter-narrative combines ideas about indigenous technical knowledge and customary institutions, including common property management rules; it points to recent studies showing the high productivity of extensive nomadic pastoralism, and the excellent adaptations farmers and herders in the drylands have made to the vagaries of dynamic, event-driven ecosystems.[12] Such a counter-narrative is a much more accurate and useful construct about what is happening in the drylands than the desertification narrative, and deserves to replace it. There are indeed many signs that this counter-narrative is taking over in the aid bureaucracies, although it has not yet got far with governments.

Such a new narrative is in part a serious attempt to engage with the views and experience of the herders and farmers who actually make the everyday decisions about land use. So far, the counter-narrative has largely been driven by researchers and by the increasing activism of dryland inhabitants. Researchers, who come out of the story of desertification with tattered banners, have a particular responsibility this time to get the science right and to ensure that the policy outcomes reflect a more just and efficient distribution of rights and responsibilities.

---

[12] See for example: Behnke *et al.* (1993), Scoones (1995), Sandford (1983), Swift (1988).

5

Wildlife, Pastoralists & Science

Debates concerning Mkomazi Game Reserve, Tanzania

DANIEL BROCKINGTON
& KATHERINE HOMEWOOD

Introduction[1]

For many years, mainstream views have held that African pastoralists overgraze; that their husbandry increases soil erosion and causes thicket encroachment on grassland. Pastoralists' stocking rates have been held to exceed the ecological carrying capacity of land, making production unsustainable and offtake per animal sub-optimal. Moreover their land tenure practices have been seen as discouraging private investment and encouraging higher stocking rates in a classic 'tragedy of the commons' (Hardin, 1968). It has been suggested that pastoralists exploit an area, over-exploit it and move on. Many observers argue this might have been possible when there was room to move but now it is anachronistic. Without a widespread change in lifestyle, pastoralists are also perceived to be a potential threat to wildlife.

This view of pastoralists' impact on the environment is still manifest in popular as well as scientific literature (for the latter see Sinclair and Fryxell, 1985). It is sustained by more than a set of ecological theories or social analyses put forward by natural scientists and anthropologists. The reasons behind the emergence, evolution and strength of the received wisdom lie in history, in views people hold of the past, and in past peoples' interpretation of their present. Accordingly, challenges to the received wisdom have done more than question the mainstream theories in natural science, economics and anthropology. It has been necessary to re-assess the history of pastoral interaction with other groups and to re-examine the history of those groups' views of

[1] This chapter draws on research in progress and funded by ESCOR of the Overseas Development Administration. The authors also wish to acknowledge the support of the Central Research Fund of London University, the Graduate School of University College London, the Royal Anthropological Institute's Emslie Horniman Fund and the Kathleen and Margery Elliot Trust. The authors wish to thank Hilda Kiwasila, Sian Sullivan and Emmanuel de Merode for their comments and advice.

pastoralists. On the one hand, it is possible to reinterpret our images of Africa and our understanding of the continent's past so that once-mainstream views about pastoralists become less self-evident. Or, on the other, it is suggested that these views might reflect different observers' desires and political agendas as much as, or even instead of, any 'real' position. These reinterpretations corroborate changing views of pastoral societies' social organization and environmental management skills and of the ecological theories of rangeland vegetation change and impact of livestock grazing.

This chapter is concerned with degradation narratives in Mkomazi Game Reserve, with the role of the received wisdom in scientific literature in creating and endorsing these narratives, and with the potential role of science in challenging that wisdom. The origins of the received wisdom, outlined above, are discussed, and recent challenges to it summarized. A framework for comparing the relevance of old and new ideas for Mkomazi is proposed. It is argued that natural science provides methods to test the applicability to Mkomazi of both the old view and the new alternative. However, anti-pastoralist beliefs are far stronger than any theories invoked to support them. Even if tested hypotheses demonstrate that new ideas are more appropriate there are still many reasons to suggest that the old policies will persist.

Origins of the received wisdom

The received wisdom has roots deep in the experience, economic policies and political interests of governments and pressure groups and, ultimately, in values strong in Western culture and collective consciousness. By tracing them and showing their potential weaknesses it is possible to demonstrate how this apparently strong position may be flawed. Scientific theory is not refuted by dividing it from its supporting historical edifice and demonstrating flaws in the latter. However the process does raise questions about the theory itself so that the challenges outlined later are more readily appreciated.

One pillar of the received wisdom is that East Africa in its 'natural' state is an unpopulated wilderness. The validity of this is now questioned. The images of Africa that late nineteenth-century explorers and administrators relayed to Europe and which gave rise to this impression are argued by some to be atypical. The early visitors encountered a land ravaged by epizootic epidemics, drought, and famine (Pankhurst and Johnson, 1988; Waller, 1988; Anderson, 1988). Bell (1987) has argued that the population of Africa as a whole was anomalously low by the time Europeans arrived. He suggests that before the 1700s Africa's rural population was quite high, based on successful indigenous land use systems and insulated from the disease pools of other continents (Crosby, 1986; McNeill, 1976; Hall, 1990; Ford,

1971; Wilmsen, 1989). Political conflicts, the slave trade, greater communication and migration with their greater attendant disease risk, all combined to cause Africa's populations to crash. Bell cites Ranesford (1983) who claims that the population of the Belgian Congo fell from 40 million in 1880 to 9.25 million in 1933. Ford (1971) has suggested that the population of Nigeria fell from 25 million in the last century to 9 million early this century. In East Africa Waller (1988) has reconstructed the effects of nineteenth-century events on the Maasai, who were decimated and dispersed by wars, famine and diseases in the 1880s and 1890s. Thus it can be argued that Africa's population was temporarily very low at the turn of the century, so that early colonists gained a misleading impression of wilderness. These initial, extreme and unrepresentative images of Africa helped to give it a special place in the European psyche and outsiders have sought to preserve what they first found. Anderson and Grove (1987) have described Africa as providing an escape from an increasingly crowded Europe. Eastern and Southern Africa in particular offered the possibility of encountering wilderness and recovering harmony with nature. The existence of wilderness, and the possibility of an alternative to the Western lifestyle has continued to be highly valued by visitors and observers from industrial nations. Threats to such wildernesses alarm them; indeed the very presence of people can be considered 'polluting'.

Africa's special role and image in European minds is, however, not sufficient to explain the depth of anti-pastoralist feeling in East Africa. Pastoralism is remarkable for the hostility with which it is viewed by governments, convinced of herders' environmentally damaging land use and uneconomic management practices. Similar views have pervaded development organizations (Sandford, 1983; Homewood, 1992a). The origins of these views can be traced to early anxieties and mistakes of the colonial era, while their persistence has been compounded by the continual frustration of policies concerning pastoralists. It is increasingly argued that the principal early errors lay in misunderstanding the nature of pastoral societies. For example, Sobania (1988; 1990) has argued that early colonial administrators' view of North Kenyan society and ecology did not match reality. The Ariaal and Rendille people were required to occupy separate restricted areas, partly because it was thought that what were perceived to be discrete tribes should not mix, when in fact ethnic boundaries were fluid. Attempts were made to improve range management by developing water supplies in ignorance of local society and ecology. The failure and rangeland degradation that resulted were blamed on the pastoralists. Similarly, an analysis of the relationship between the Maasai and the Kenyan colonial authorities (Collet, 1987) shows that from as early as 1893 the Maasai were viewed with antipathy. They were characterized as warlike, lazy and inefficient, preying on the groups around them and wasting valuable land by refusing to cultivate. The British justified land appropriation by claiming that it put the land to good use and aided the civilization of the Maasai through the 'hard honest work' of cultivation. Development was to be achieved by encouraging agriculture and agriculturalists on Maasai lands.

Nevertheless, the antipathy was not universally felt. Waller (1976) has traced the convergence of interests between British colonials aiming to 'pacify' more northern tribes in Kenya, and the Maasai who were able to rebuild their herds by participating in punitive raids. Many officials, predominantly local administrators, felt that Maasai lands should be protected from invading settlers of other groups (Waller, 1993). Pastoralists always had some supporters – such as Fosbrooke (1948) in Tanzania – but their arguments have been perennially overridden by governments frustrated by their unprofitable interactions with pastoralists (Sandford, 1983; Lane, 1991).

When the northern Maasai were moved from their lands before the First World War it was initially recognized that they did not harm wildlife, and that animals were relatively safe on their lands (Collet, 1987). However it was increasingly feared that wildlife would be progressively excluded from Maasai lands as their stock increased. This view gained strength from the belief that the Maasai were irrational herders, accumulating stock beyond their needs (Herskovits, 1926) until the growing herds inevitably displaced wildlife. In the 1930s and 1940s anxieties over the effects of herd increases on wildlife were compounded by fears for the environment. However, Anderson (1984) demonstrates that environmental concerns in Kenya were not founded on evidence of *in situ* environmental degradation alone. Fears arose partly from erosion events elsewhere and did not necessarily reflect environmental damage in East Africa. Anderson argues that in the 1930s, these fears were exploited to further political ends. Sparked off and reinforced by awareness of the Dust Bowl in America, they were used to justify land expropriation for, and subsidies to, settler farming during a recession.

The impact of the Dust Bowl on the collective consciousness of agricultural officers was almost as great as its impact upon the prairies of North America. Because of the damage experienced there, range scientists became extremely concerned about the amount of vegetation cover. Pasture condition was (erroneously) measured in terms of cover quantity, regardless of its utility as forage (Smith, 1988). Recognition of the erosion problem and the need for vegetation cover and other efforts to contain it became prerequisites for a successful career. The frequent occurrence of bare ground on East African pastoral lands meant that as American ideas and training filtered through to East Africa, the assumption that there were dangerous levels of soil erosion on pastoral land became conventional wisdom.

Demographic pressure in the African Reserves of Kenya led to visible degradation and 'the spectre of erosion galloping out of the Reserves and into the White Highlands seemed all too real' (Anderson, 1984: 329). The coincidence of these concerns at home, fed by a period of low rainfall from 1926 to 1935 and disasters abroad, raised fears that the area was drying up and that somehow human practices were causing the area to desiccate. At the time European settlers had to convince a sceptical government that they should have more land despite the high cost and inefficiency of their farming. The settlers argued that the apparent erosion threat posed by African occupancy of land meant that a responsible government had to put more land into settler hands if future resources were to be safeguarded. Erosion fears were amplified to meet political ends, and the resulting image did not necessarily match the reality of change in land quality.

The colonial view that pastoralists are damaging to the environment and threatening to wildlife has been perpetuated and reinforced in post-independence Africa. Numerous efforts made to 'develop' pastoralists and incorporate them into the modern economy have achieved little (ILCA, 1980; Galaty *et al.*, 1981; Sandford, 1983). The failure to extract a profitable surplus from pastoralists is largely rooted in misunderstanding of their motives and social organization. However it has only served to reinforce the view that pastoralists are irrational people, unresponsive to government interests and development efforts (Lane, 1991; Galaty and Bonte, 1991), whose livestock-accumulating tendencies threaten their own environment (Sandford, 1983). This is particularly true where pastoralists' needs appear to conflict with those of the tourist industry. Action here can be particularly dramatic because of the urgent need to preserve threatened species, and because of the monetary value of the tourist trade. Because they are assumed to be dangerous to wildlife, pastoralists are excluded from parks and reserves in order to protect the resources there. Initial prejudice then becomes self-fulfilling prophecy as conflict mounts over protected resources. This is exemplified in the spearing of rhino and elephants by pastoralists in Amboseli, which followed worsening relations between the park authorities and the Maasai (Lindsay, 1987; Western, 1982a).

Challenges to the received wisdom

Colonial and post-colonial antipathy towards pastoralists has been supported by theories of range ecology and economics which purport to explain how pastoralists' stocking rates and land tenure arrangements combine to cause environmental degradation in the so-called 'tragedy of the commons' (Hardin, 1968). Analysis of the complex origins of and support for these theories cannot alone invalidate them. If pastoralists today have too many cattle and cause overgrazing, and if animal husbandry threatens wildlife, then the established scientific arguments stand. However there are many reasons to suggest that the relationship between cattle numbers and grass cover, livestock and wildlife, and pastoralists and their environment, are more complex than once thought. Pastoralists and wildlife have co-existed in East Africa for thousands of years, and present populations of wildlife are concentrated on pastoralist rangelands (Peden, 1987; Homewood and Rodgers, 1987, 1991). A number studies of wildlife and livestock ecology suggest that the two may compete for resources yet still coexist in the long term (Peden, 1987; Western, 1982b; Homewood and Rodgers, 1991). New ecological theories suggest that many pastoralist practices, until recently seen as leading to environmental deterioration, are in fact sustainable (Homewood and Rodgers, 1987; Behnke and Scoones, 1993).

New thinking in range ecology suggests that because of the resilience of savanna vegetation communities, they are not prone to collapse even under heavy grazing pressure. In semi-arid and arid environments it is rainfall rather than levels of grazing that determines productivity. The deterministic successional theories that once dominated rangeland ecology are now largely discredited (Smith, 1988). Change in species composition is now thought unlikely to reflect progressive change between successional stages. Instead, species composition is now believed to reflect complex stochastic effects of climate interacting with rare species-recruitment events, the initial abundance of particular species populations, the demography and biology of individual species, and grazing and fire regimes (Noble, 1986). These 'disequilibrium' models of vegetation change now allow for several pathways of change and for reversals of change (Westoby *et al.*, 1989). The possibility of reversible change on rangeland means that not all 'degradation' is necessarily permanent.

Permanent change is more likely if soils are seriously affected. However, estimates of soil erosion on rangelands under pastoralist regimes are being reduced (Abel and Blaikie, 1989; Abel, 1993; Biot, 1993; Swift, this volume), and the seriousness of that which does occur is being questioned (see also Stocking, this volume). Bare ground does occur, but is not necessarily as dangerous as agricultural officers once perceived. It is increasingly argued that pastoralists' presence has a positive impact on the ecosystem, and that people could be considered a positive, natural component of the environment (Homewood, 1992b; 1994; Mwalyosi, 1992). Herders may improve pasture by creating nutrient rich patches on old boma sites, for example, while fire regimes used by pastoralists can increase the nutrient status of pastures (Scholes and Walker, 1993). This is not to argue that overgrazing never occurs, since pastoralists themselves recognize that it does. But it does suggest that sustainable levels of grazing may be much higher than was once thought, and that the tendency to attribute changes in the environment to grazing needs to be treated with caution. Allegations of irreversible environmental change, in particular, appear less plausible.

There are also grounds for believing that local communities' resource use is well organized and controlled. Tragedies of the commons (Hardin, 1968) are not a necessary consequence of 'communal' ownership. This is to confuse resources held in common, which may be managed to prevent selfish use, with open access. Alternative theories of communal resource management and pastoral production systems suggest that pastoral people act to ensure that grazing is controlled (Berkes, 1989; Bromley and Cernea, 1989; Lane and Swift, 1989; McCabe, 1990; Homewood and Rodgers, 1991; Potkanski, forthcoming). This comes as part of an established and growing body of knowledge that demonstrates the depth of knowledge that local resource users have about their environments and the wisdom with which they use them. It is not suggested that local practices are perfect: they can cause damage, but experience shows that improved understanding of pastoralists makes practices appear more reasonable. Thus Herskovits' (1926) 'cattle complex' was not entirely wrong; cattle do occupy a special place in pastoralist mentality that reason and need cannot explain (Galaty and Bonte, 1991). However these cultural values are balanced by expert knowledge of range condition, herd and risk management and market forces. For example, Scoones (1989b) has demonstrated the high accuracy and precision of botanical and edaphic knowledge amongst Zimbabwean farmers, and Sandford (1982b) has shown that the low stocking rate of commercial meat ranches carries considerable opportunity costs in an arid environment with fluctuating rainfall. Moreover, comparisons of pastoralist productivity with that of ranches have tended to emphasize the goals and achievements of the latter while failing to include important aspects of the former (Behnke, 1985). Once the need to conserve herds against the threat of uncontrollable risks that beset rangelands are taken into account, pastoralists' herds do not seem as excessive as previously thought (Coughenour *et al.*, 1985).

Finally there are beginning to be common challenges and alternatives to the perceived need to protect wildlife by removing people. These dovetail with broader shifts in approaches to conservation, which lay greater emphasis on reconciling it with the needs of inhabitants. The situation in Amboseli outlined above, for example, has changed. Agreements have been reached between park authorities and the Maasai, and some argue that this can increase controls over poachers (Western, 1994). In Ngorongoro, Tanzania, it is argued that wildlife competes successfully with livestock, and that the pastoralists rather than the animals are on the retreat (Homewood and Rodgers, 1991). Outside protected areas, extension work, adult education, and projects channelling wildlife revenues to local communities are encouraging wildlife to be viewed as a valuable resource and enhancing the possibilities of cooperation between local communities and those in charge of wildlife (Berger, 1993; Swanson and Barbier, 1992; Lane *et al.*, 1994).

Mkomazi

Notwithstanding these recent challenges, received wisdom concerning pastoralists' practices and their relationship with wildlife is still manifest in the policy towards pastoralists in Mkomazi Game Reserve (MGR) and in scientific work written and planned for the area. MGR occupies an area of rangeland on the north-east border of Tanzania (see Figure 5.1), contiguous with Tsavo National Park in Kenya. It has an average rainfall of approximately 600 mm/year, although this is subject to considerable fluctuations. It was gazetted as a game reserve in 1951 and pastoralists then resident were allowed to remain (Watson, 1991; see Table 5.1). In the late 1970s and 1980s, however, concern mounted over environmental changes perceived to be caused by the increasingly numerous pastoralists. They were asked to leave and efforts were made to find suitable alternative areas. In 1988 they were evicted (Mduma, 1988) and appear to have suffered considerably from being denied access to reserve resources (Mustafa, 1993).

Earlier writings concentrated on documenting the natural history of the area (Ansty, 1955; Watson *et al.*, 1969). Later Harris and Fowler (1975) modelled the herbivore population dynamics of the ecosystem to establish patterns of vegetation change consequent upon different human and elephant populations. This was done in order to emphasize and demonstrate the highly dynamic processes that create the state of the Mkomazi ecosystem. They saw that active management would be required in order to maintain biological systems at some reasonably constant composition as was generally realized for savanna ecosystems at the time. This represented a move away from considering the landscape of East Africa to be constant, although the thinking still emphasized management in order to preserve a single stable state, rather than management that accepts dynamic states (cf. Pellew, 1983; Robertson, 1987).

Harris and Fowler's work is of interest in the light of current ideas of disequilibrium theory and of state-and-transition models of rangeland ecology. However, it is not clear that their contribution has had as much impact as later, more speculative and journalistic contributions by natural scientists to the degradation narrative on Mkomazi. The recent contributions of natural scientists to the environmental narrative concerning MGR have been based on conventional wisdom over pastoralist-induced degradation. They have emphasized exclusive fortress conservation practices and endorsed the exclusion of pastoralists from Mkomazi on ecological grounds. Sembony (1988) documented local disquiet among administrators at the gradual loss of the Reserve to encroaching herders. Mduma aroused concern by stating that pastoralist human and livestock populations were increasing with 'adverse effects' (1988: 17). Evidence of wildlife or environmental decline caused by this pastoral impact was not given, but the author hints that unless the

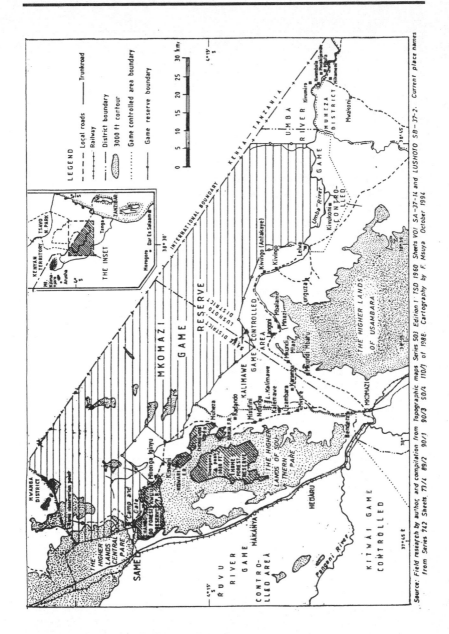

Figure 5.1   Location of Mkomazi Game Reserve
Source: Field research by Hilda Kiwasila and compilation from topographic map Series 503
Edition 1 TSD 1960 Sheets VOI SA-37–14 and LUSHOTO SB-37–2. Current place names
from Series 742 Sheets 73/4 89/2 90/1 90/3 110/1 of 1988. Cartography by F. Msuya,
Institute of Resource Assessment, Dar es Salaam, October 1994.

Table 5.1   The History of Mkomazi Game Reserve

1951	Reserve established by British Colonial Authorities, to replace the Ruvu Game Reserve which was thought too degraded for wildlife.
1952	List of Parakuyo pastoralists allowed to live in the area compiled in accordance with the Faunal Conservation Ordinance.
1957	Boundaries altered: Kalimwe area excised for grazing and cultivation. Dindera Dam built.
1963	List of legal residents corrected (this probably means increased).
1965	Boundaries altered: Igoma area excised for grazing and cultivation.
1967	Kavateta and Ngurunga Dams built.
1968	List of legal residents increased.
1971	List of legal residents increased in January and again in November.
1973	Split into the Mkomazi and Umba Game Reserves under Kilimanjaro and Tanga Regions respectively.
1974	Boundaries and management regulations stipulated in the Wildlife Conservation Act of 1974 which replaced the Faunal Conservation Ordinance. Residence was allowed under the terms of this act.
1976	The manager of the Mkomazi Game Reserve writes to the pastoralists staying there saying that the permits previously given were revoked due to alleged degradation consequent upon overgrazing.
1981	Pastoralists meet with District Commissioner and Party representatives to discuss alternative areas in which the pastoralists have been told they can settle.
1987	The Regional Office writes to pastoralists in the Mkomazi and Umba Game Reserves cancelling all previous permits to live there and ordering them to leave.
1988	Pastoralists evicted.

Reserve is properly protected then all the animals face extinction (1988: 18). He also reports that the problem of land degradation has become apparent, though no evidence for this is given, nor is the link between environmental change and pastoral activities demonstrated. Nevertheless Mduma recommends that all pastoralists be moved if the Reserve is to thrive. Watson similarly observes that Mkomazi was 'nearly grazed off the map' (1991: 1), although again supporting evidence is not provided. The concerns that the pastoralists were causing overgrazing may have been based on an official report, but no such document is cited by any of the authors reported here.

Planned research organized under the auspices of the Royal Geographical Society (RGS) reflects the strength of these established views. The following statements are drawn from the Scientific Report for the Mkomazi Ecological Research Programme 1994–96:

The Maasai pastoralists with their herds of domestic stock, have steadily increased their presence in some areas over the last 20 years. As a result the vegetation of the western and central areas of the MGR have undergone considerable changes through the effects of feeding pressure from domestic animals and especially fire. (RGS, 1994: 3)

Since the removal of pastoralists from Mkomazi, the number of elephants observed has increased from 20 or 30 to well over 600. This is a sure indicator of the importance of maintaining a low level of human disturbance now and in the future. (RGS, 1994: 4)

The received wisdom also influences the policy of an NGO, the George Adamson Wildlife Trust, which is currently working on community outreach programmes around Mkomazi. The exclusion of local inhabitants by the Reserve management carries logical implications for community outreach. If people so affect the ecosystem that they have to be excluded, then they have to be taught the reasons for exclusion and how to cope with it. The main efforts of the programme are accordingly directed at improving people's sense of responsibility for the environment, raising awareness of environmental problems, and curbing the worst excesses of current practices by reorganizing land tenure, and improving resource use. The programme's views of local resource use can be inferred from these plans, and are explicit in the following statements, drawn from the manifesto of the Programme:

Those who depend upon their surroundings for their living are not in a position to take care of their environment. (Anon, n.d.: 2)

Environmental degradation is . . . also connected with unawareness, irresponsible attitudes, a lack of interest in the future world and also a lack of interest in the environment of their labour area . . . everybody is trying to escape from his responsibility to the environment. (Anon, n.d.: 6)

The role of science in challenging received wisdom

Recent literature dealing with the environment of Mkomazi, and the capabilities of Tanzanians resident near it, is clearly largely at odds with many of the new theories and empirical findings that have emerged over the last decade from studies of comparable systems elsewhere. The future contribution of natural science in resolving the impasse lies in the fact that the theories of both sides are testable. Explicit and implicit in the literature are refutable ideas on environmental change and resource use in Mkomazi. Where a paradigm begins to change, scientific investigations provide a powerful approach whereby competing theories may be tested in a more objective way. They can establish a framework of competing hypotheses and methods to test them in order to reach an objective understanding of what is going on in the Mkomazi ecosystem. Such hypotheses concerning the environment's workings and the management of Mkomazi can be extracted from the recent literature. Taken together, the hypotheses summarized in Table 5.2 represent the

Table 5.2   Contrasting Perceptions of Mkomazi

	Received Wisdom	Alternative View
Environmental Change	1. Pastoralists damage the ecosystem by overstocking. – Soil erosion increases. – Rangelands vegetation becomes less palatable. – Bush invades pasture.	1. Pastoralists do not damage the environment by overstocking. – Soil erosion does not increase to dangerous levels. – Vegetation dynamics are not driven by stocking rates. – Pastoralists make a positive impact upon the spatial nutrient concentrations of the area, leaving nutritious grazing on old *boma* sites.
Wildlife	2. Pastoralists compete for resources with wildlife to the exclusion of the latter. – Pastoralists physically exclude wildlife from water sources and use up water, not leaving enough for wildlife. – Pastoralists exclude wildlife from good pastures and use up grazing.	2. Pastoralists and wildlife do not compete to the exclusion of the latter. – Competition over water and grazing is not sufficient to threaten the viability of the Reserve's wildlife populations. – Pastoralists are excluded from some sites by the threat of diseases held or transmitted by wildlife.
Burning	3. Local burning practices are turning woodland into grassland and decreasing the Reserve's biodiversity.	3. Changes in vegetation now observed are due to changing intensity and timing of the burns. – Exclusion of pastoralists and their cattle has left more vegetation to burn causing hotter fires.
Elephants	4. Exclusion of pastoralists is directly responsible for the present increase in elephant numbers now observed in Mkomazi.	4. Increases in the elephant population in Mkomazi reflect improved poaching controls in Tsavo National Park, not the absence of pastoralists in the Reserve.
Hunting	5. The local population's hunting activity threatens wildlife populations.	5. Local hunting activities are not endangering wild ungulate populations.
Management	6. Local communities' resource use is not well organized. – Either: they are unable to act collectively of their own accord to prevent degradation resulting from their use of resources. – Or: attempts which they do organize are ineffective.	6. Local communities do organize their resource use. – Levels and extent of use are agreed and negotiated. – Sanctions are available to punish those who ignore these agreements.

Table 5.2   (*Continued*)

	Received Wisdom	Alternative View
Land Tenure	7. The land tenure arrangements inevitably encourage unsustainable use. – Land tenure arrangements should change from communal to private wherever possible.	7. Use of communal land can be controlled. – Changing the form of tenure will not prevent degradation.
Environmental Awareness	8. The health of the environment does not feature highly in local peoples' conciousness. Degradation does not concern them sufficiently to provoke preventive action.	8. Local people are aware of environmental problems. – They act to alleviate them.

spectrum of standpoints on Mkomazi, although they do not represent the opinions of any single person or group.

Kuhn (1962) suggested that knowledge only changes when the protagonists of the old ideas give up defending them from the attacks of other zealous scientists, and that orthodoxies are as enduring as the energy and mental agility of those holding them. But Gellner (1992) makes it clear that for all the cultural relativity of Kuhnian paradigms, natural science has powerful tools that can be used to test theories in an objective way. This is not to suggest that science is neutral. Natural scientists today are as much a product of their own cultures as early colonists and the historical record behind the received view of pastoral impacts upon the environment demonstrates that science can be used to other ends. However, the tools of natural science may be used in Mkomazi to attempt to distinguish between the implications of two radically different sets of views to examine the extent to which observed data are compatible with one or other interpretation.

It is nevertheless possible for paradigm change in natural science to have little impact on policy. Indeed, as the received wisdom in the Mkomazi case is more than a set of scientific theories, it will take more than refutation within science to challenge it. First there is the difficulty of administering both pastoralists and wildlife together. Mkomazi was co-managed from its inception until 1988, and the current exclusion of pastoralists is a response to the perceived failure of joint use. Ecologists thus cannot simply proclaim pastoralism to be compatible with wildlife; they have to demonstrate how shared use would work in practical terms. Second, the sense of urgency in African conservation, and the pressure to 'do something' for African wildlife 'before it is too late', means that the slow, cautious methods of natural science are not attractive to reserve managers. Rare species are so important and the tourist trade so valuable that any threats are taken seriously and dealt with decisively. Conservationists consider that there cannot be a 'wait and see' attitude where endangered species and a valuable industry are concerned. This need to make quick decisions, and stick to them, means that science cannot answer many of the problems that reserve managers face on appropriate timescales.

Finally and fundamentally, the conflict surrounding Mkomazi is a conflict of values between ideas of how the environment should be used and what the landscape should look like. Ecological theories may be invoked to support them but the values that exclude people from wilderness are not so much theory-dependent as lying far deeper in culture and consciousness. The idea that wild Africa should be 'people-less' and that wildernesses should be separated and preserved from people are sacred absolutes, revered by generations of conservationists, and advocated with a passion that scientific refutation cannot touch. The fear of people, and of their present and unknown future impact, means that the language of paradigm change in natural science may be inappropriate when discussing change in management policy. The debate is, in effect, being played out on a different stage. Similarly, newly emerging ideas of compatibility between conservation and indigenous communities are not based on new data alone. Under-pinning them is the belief that the needs of people matter as much as, if not more than, those of animals, and that it is unethical for Africans to be excluded from resources so that wealthier European tourists can use them. Here is a conflict of values. Ultimately new theories purporting to demonstrate compatibility of people and wildlife will have difficulty persuading those who believe *a priori* that people should not be there.

6

Rethinking the Forest-Savanna Mosaic

Colonial Science & its Relics in West Africa

JAMES FAIRHEAD
& MELISSA LEACH

Introduction[1]

Since the 1890s, scientists and policy-makers have considered the patches of dense, semi-deciduous forest found scattered in the savannas on the northern margins of Guinée's forest zone to be relics of a previously more extensive forest cover. There are about 800 such 'forest islands' in Kissidougou prefecture alone, most concealing at their centre a clearing containing one of the prefecture's villages. The existence of forest patches amidst savanna has suggested the penetration of savannas southwards into the forest zone as a result of vegetation destruction by farmers. A century after its first elaboration, this interpretation of a landscape half empty of forests continues to drive repressive policies designed to reform inhabitants' land-use practices.

This received wisdom concerning the forest-savanna mosaic refers to historical processes, but is not founded on historical data. When the historical record is consulted, it suggests a strongly contrasting reading of vegetation change; one of a landscape 'half full' of forests. In this chapter we examine how the agro-ecological knowledge and experience of local inhabitants – invalidated within colonial visions – explain plausibly the demonstrable history of Guinée's forest-savanna mosaic. Furthermore, they also reinterpret evidence once presented in support of arguments concerning its degradation. We go on to explore how the scientific conviction that forest islands were relics has been incorporated into administrative canon and sustained there, considering

[1] This chapter has been equally co-authored by Melissa Leach and James Fairhead, and draws on research carried out with the assistance of Dominique Millimouno and Marie Kamano. The research was funded by ESCOR of the Overseas Development Administration, whom we gratefully thank; opinions represented here are, however, the authors' own, not those of the ODA. Many thanks are also due to the villagers in Kissidougou prefecture with whom we worked and to our Guinean collaborating institutions: Projet de Développement Rural Intégré de Kissidougou (DERIK), Direction National des Forêts et de la Chasse (DNFC), and the Direction National de la Recherche Scientifique (DNRS).

particularly how institutional, financial and socio-political structures have shaped environmental policy. In this light, rethinking the forest-savanna mosaic not only carries very different implications for policy, but also suggests the need for a more fundamental challenge to these structures.

A century of relic visions

The relic vision of Kissidougou's forest-savanna mosaic is as old as the first French presence there. In 1893, the prefecture's first military administrator, Captain Valentin, mentioned in his first report on the area that:

> The soil of the valleys has a more or less thick humus bed. The humus derives, as one understands it, from the immense forests which cover a large part of the soil, and which covered it entirely at a period relatively little distant from our own.[2]

Valentin's analysis was consistent with his broader view of the natural prosperity of this zone. For the early colonists of French West Africa, extensive forests epitomized an environment replete with agricultural potential and economically exploitable wild products, in contrast with the drier zones further north. Valentin based his conviction that the zone's earlier vegetation was full forest not only on its fertile soils and large populations, but also on the presence of oil palms (*Elaeis guineensis*), which he assumed to indicate where forest had once been.

Valentin's 1893 reading of the landscape has been repeated throughout the twentieth century, although observers have commonly updated the presumed period of intact forest cover. When the botanist responsible for Kissidougou's forest policy in the 1940s wrote up his data, for example, he suggested that extensive forest cover had existed in 1893: 'The information obtained from the oldest inhabitants confirms what we supposed. The whole region was covered with forests around 75 years ago' (Adam, 1968: 926). Adam himself described the landscape which he saw as 'oases of equatorial vegetation in the middle of savanna burned by the sun and fire . . . all in regression' (1948: 22). Today's national forestry plan also sees forest in regression: 'the quasi general opinion is that . . . Kissidougou will soon be no more than a vast poor savanna, the islands and gallery forests still present at risk of being rapidly destroyed' (République de Guinée, 1988: 31). Several modern studies informing environmental and rural development projects consider extensive forest cover to have been lost within the last 50 years. One reported that 'around 1945, the forest, according to the elders, reached a limit 30 km north of Kissidougou town. Today , its

[2] Valentin, 'Rapport sur la Résidence du Kissi', 1893, Archives Nationales de Senegal, Dakar, 1G188.

northern limit is found at the level of Guekedou-Macenta, thus having retreated about 100 km . . . This deforestation is essentially the result of human action' (Ponsart-Dureau, 1986: 910). A consultant examining fire policy for Kissidougou's integrated rural development project argues equally that in

> . . . the green belts which surround the villages, one finds the relics of original primary forests. The value of these biotypes in the heart of a nearly 100 per cent degraded environment is inestimable. One finds no individual of [characteristic savanna species] more than 35 years old . . . supporting the thesis that the site has burned systematically only since then. (Green, 1991: 10–11)

While this view that forest loss is recent is shared by policy-makers dealing with Kissidougou, not all scientists have agreed. Some have found indicators for the presence of savanna over many centuries. Notably, for example, Schnell's (1952) observations of old baobabs suggested to him that Kissidougou had been deforested 'early'. All observers of Kissidougou have nevertheless considered the landscape in terms of a one-way deforestation process, and have assumed that local land-use is responsible, being inherently destructive. Shifting cultivation of upland rice and fire-setting in farming and hunting were deemed responsible for the progressive savannization of forest. Elsewhere in West Africa, isolated ecological studies have argued that the forest-savanna mosaic is a more stable vegetation pattern, where the presence of forest is associated with the distribution of particular soils more favourable for its establishment (e.g. Moss and Morgan, 1977; Avenard *et al.*, 1974). These still consider the forest patches to be natural formations, although they depart from the view that they are necessarily relics, thus presenting people less as destructive than benign in their influence on forest cover.

Re-viewing the forest-savanna mosaic from longitudinal evidence and local agro-ecology

When historical data are used to trace vegetation change, a very different reading of Kissidougou's forest-savanna mosaic emerges.[3] First, we compared modern (1989–92) air photographic and satellite images with air photographs from 1982, 1979 and 1952. This comparison shows that in many zones, the areas of forest and savanna vegetation have remained remarkably stable during the 40-year period which today's policy-makers consider to have been the most degrading. Where changes are discernible, these predominantly involve *increases* in forest area. In the northern part of the prefecture, there is evidence of

---

[3] For further details of this historical analysis, see Leach and Fairhead (1994) and Fairhead and Leach (1996).

additional patches of forest vegetation, and of increasing savanna tree density. In the south, large areas covered by savanna in the 1950s have ceded to a forest thicket or bush fallow vegetation. Within these broad shifts, more micro-level dynamics are also evident: forest islands have diminished or disappeared in some places, but enlarged and appeared in others.

For earlier periods, we compiled a picture of vegetation from landscape descriptions and maps found in archives and explorers' reports. While they leave room for ambiguity, those from the turn of the century (1893–1914) nevertheless clearly falsify assertions of more generalized forest cover. Rather, they suggest not only that the areas of many forests may well have grown from this earlier date, but also that many savannas were, at the turn of the century, almost treeless. Earlier documentary sources from the 1780s–1860s certainly do not suggest the existence of forest cover. Both Harrison, travelling to Kissi areas *c.* 1785 (see Ludlum, 1808; Hair, 1962), and Seymour in Toma country south east of Kissi, in 1858, describe short grass savannas and an absolute scarcity of trees in places which now support extensive dense humid forest. This early picture of more extensive grass savannas is also consistent with current evidence concerning the region's longer-term climatic history, with drier conditions than today probably prevailing between c.1300 and 1850 (Nicholson, 1979).

For a third sort of evidence, we developed and used methods for collecting accurate oral information concerning past vegetation from Kissi and Kuranko inhabitants. In the south and south-east, changes in everyday resource use confirm that grass and sparse shrub savanna over large areas has ceded entirely to secondary forest thicket. The introduction of tree felling into agricultural operations, the increased availability of preferred fuelwood species, the changes in the materials used in roofing and thatching, and the changing patterns of bird nesting, distribution and control leave no doubt (Fairhead and Leach, 1996). In more northerly zones, villagers describe changes in the palm oil economy which indicate increases in the density of oil palms during the present century. Most strikingly, oral accounts reveal that village forest islands are generally formed through human settlement and management. In 71 per cent of the 38 villages we specifically investigated, elders recounted how their ancestors founded settlements in savanna and gradually encouraged the growth of forest around them. While such histories might reasonably be considered as politically conditioned fiction, stressing a founder's firstcomer status in creating prosperity in a hitherto empty landscape (cf. Dupré, 1991), we can be more confident in them because the processes they describe can be observed today in the formation of forest around more recently established settlements.

Unsurprisingly, villagers' own explanations of the vegetational effects of their land use are consistent with this vegetation history.

Demonstrable patterns of vegetation change can largely be accounted for through an understanding of local agro-ecological management practices, their vegetational impact, and their changing application under shifting economic, political and demographic conditions.

People value forest islands around their villages for numerous reasons, whether for the protection they provide against bush fires, high winds and excessive heat; for convenient sources of forest products or for suitable microclimatic conditions for tree crops. Historically, forests reinforced fortification, and parts of them continue to house men's and women's initiation and ritual activities. Inhabitants therefore encourage forest island development more or less deliberately, principally by altering fire and soil conditions so as to favour forest regeneration (Leach and Fairhead, 1993; 1994). Collecting thatch and tethering cattle on the village margins remove flammable grasses and help create a fire-break. When necessary, this can be supplemented by a targetted burn in vegetation on the settlement margins early in the dry season to reduce the effects of later, more damaging fires. Gardening, villagers assert, tends to create soil structure and water relations favourable for tree establishment. Sometimes settlements are founded on old garden sites, or their margins are deliberately gardened for a limited period to help forest establishment. The village-edge soils are also fertilized by the excreta of domestic animals and people, and by hearth ash and other wastes scattered there. For all these reasons, savannas on settlement margins tend gradually to develop dense semi-deciduous moist forest vegetation. Initial fire-resistant tree species gradually cede to more fire-intolerant ones, pioneer forest species cede to those typical of later forest successional stages, and biodiversity increases over time.

At times, villagers have accelerated the formation and establishment of forest islands by planting trees. Before the twentieth century, for instance, silk cotton (*Ceiba pentandra*) and other fast-growing tree species were commonly planted as part of settlement fortifications. Islands have also been modified and expanded to create tree crop plantations, whether for the kola anciently valued and traded in this region, for coffee which became an important cash crop from the 1940s, or for the fruit trees and bananas currently favoured by villagers. Villagers also enrich forest islands with useful food, medicinal or construction species, by transplanting wildlings or encouraging their growth from suckers or cuttings.

For inhabitants, then, forest islands are far from the relics of a disappearing 'nature'; instead they are strongly associated with settlement, existing because of it and its everyday activities. As a corollary, village abandonment can be associated with forest island loss, since an uninhabited island becomes more difficult to maintain and perhaps more advantageous to convert to farmland, benefitting from the super-fertility of the enriched site. Forest islands thus tend to come and go with changing population distribution and settlement patterns, in a dynamic which effectively accounts for demonstrable changes in their numbers and distribution in Kissidougou's vegetation history (Fairhead and Leach, 1996).

Between the forest islands, villagers value and use both savanna and forest (or forest fallow) for farming rice and other food crops, grazing cattle, and collecting plant products. When they use forest vegetation, villagers do not consider that they are converting it enduringly to savanna; indeed in the instances when they do perceive this as a risk – as when rice-farming in forest fallow sites in fire-vulnerable areas – they will take steps to prevent it. According to villagers, certain agricultural and livestock techniques, when practised in savanna, tend to increase its woody cover, and even assist its conversion to forest or forest fallow vegetation. This is the impact, for example, of the long cultivation sequences often used when farming peanuts, cassava and fonio, in which repeated gardening-like mounding, and the burying of organic matter over the years, improves the soil's structure, water relations and fertility in an enduring way. When abandoned, woody vegetation rapidly establishes on such sites which gain some fire protection both from the soil's improved water-holding capacity, and from the less flammable grasses which dominate post-cultivation.[4] Once 'gardened' sites can thus be associated with dense vegetation, whereas 'new', never-cultivated land tends to carry relatively open savanna. The prevalence of such gardening within local farming patterns has increased since the 1950s, linked to an increase in independent food and cash-cropping by women farmers (Leach and Fairhead, 1995), and this helps to account for the multiplication of woody areas in the landscape which has been observed. Villagers also find that intensive cattle-grazing can similarly 'deflect' savanna to forest successional vegetation, by reducing flammable grasses, manuring soil and importing seeds (cf. Boutrais 1992). Cattle numbers have tended to increase since the mid-twentieth century, recovering from the periods of pre-colonial warfare and of heavy colonial taxes, both of which decimated herds. This seems also to have contributed to woody cover increases in parts of the prefecture where cattle are abundant. In some southerly localities, however, pasture has now entirely disappeared, and cattle-keeping with it.

Fire management practices have also contributed to vegetation change. In the drier, more sparsely populated northern regions, fire cannot easily be fully prevented, and villagers have long used early-burning, and the sequencing of fire-using activities through the dry season, to limit the damage which late dry season fires can cause to fallows and property. These practices have assisted the maintenance of

---

[4] Local reasoning in this respect is supported by studies of forest-savanna ecology which show the critical role of soil surface moisture characteristics and profile heterogeneity in tree seedling survival (Moss and Morgan, 1977), and the role of cultivation and associated termite activity in improving soil texture, fertility, and water relations, increasing water supply to vegetation cover (e.g. Lal, 1987; Casta *et al.*, 1989).

the forest-savanna mosaic. In the more humid and densely populated south, fire management practices have centred on preventing running fires using fire-breaks and mass-mobilization to extinguish them. As population growth, changing land-use patterns and the climate of external regulations have increased villagers' abilities and incentives to apply these techniques, so they have become more effective in some areas, assisting the cession from savanna to forest vegetation (Fairhead and Leach, 1996).

The persistence of the deforestation orthodoxy

These increases in woody cover – demonstrable from historical data, and compatible with the experiences of many rural inhabitants – have been invisible to scientists and policy-makers, who throughout the period of their occurrence remained convinced that Kissidougou was losing its forest cover fast. How was their conviction maintained? It is clear that the colonial view of Kissidougou's landscape as degraded and degrading was partly reproduced and elaborated through the inheritance and transfer of ideas among individuals and institutions. Nevertheless, successive generations of observers as much 'rediscovered' this 'half-empty' analysis, and its endurance also needs to be examined in relation to the prevailing intellectual and social structures.

Only partial information concerning the 'genealogy' of the savannization view is available, but this nevertheless reveals important links both among scientists and between scientists and administrators – who were often one and the same individuals.[5] From the turn of the century, the botanist Chevalier made tours to collect specimens for the colony's botanical gardens, and developed the analysis of regional climate and vegetation zones. In this, he considered the transition zone as ex-forest, largely maintained as savanna by the fire-setting of inhabitants. The archives testify to the dominating influence his knowledge had on staff within Guinée's developing (and initially combined) agricultural and forestry services.[6]

Chevalier's early reading of the ex-forest landscape – like Valentin's in 1893 – did not condemn it. Indeed, in 1909 he actually identified southern Kissidougou as a model for the country:

> The province of Kissi seems to us to fulfill currently what one must endeavour to obtain in all of Guinée. The managed forests there cover a rational area and alternate with de-wooded savannas and lands reserved for crops. Thanks to these forests the rainfalls are

[5] The scientist Chevalier, for example, worked as a botanical consultant to Afrique Oriental Française (AOF) while the botanists Adam and Aubréville worked for the AOF forestry service, the latter rising to be Inspecteur Général des Eaux et Forêts des Colonies.

[6] Guinée Service de l'Agriculture et des Forêts, 'Rapport Annuel, 1913', Archives Nationales de Sénégal, 2G13(1).

regularised, the agricultural lands are maintained and the indigenous inhabitants find . . . quantities of resources which do not exist in the bush in the strict sense of the word.[7]

Nevertheless, as these remarks indicate, climatic and soil considerations already ranked among the perceived reasons for forest protection, and others influenced by Chevalier's analysis were far less optimistic about the stability of Kissidougou's landscape. Prompted by a dry period in 1914, Nicolas, the director of the agricultural research station remarked:

> Never, I believe, has a year so dry occurred in Kissidougou. I am left to say that from year to year, rain becomes more and more rare. And this I do not find extraordinary – even the contrary would astonish me – given the considerable and even total deforestation in certain parts of this region. From Kissidougou to Gueckedou, all has been cut . . . the effects of this de-wooding are disastrous; one will soon see nothing more than entirely naked blocks of granite. A region so fertile become a complete desert. Now there rests no more than a little belt of trees around each village and that is all.[8]

These early ideas concerning climate, soils and 'desertification' were gradually elaborated in more scientific terms in soil science, botany and regional ecology, especially in the work of Chevalier and culminating in the monumental work of Aubréville (1949). They were disseminated through scientific tours and conferences, including those linking French and British colonies.[9] The commonplace perceptions of administrators – such as that dry seasons and the harmattan were intensifying due to fire-induced vegetation change,[10] and that savannas were shifting southwards – thus acquired the weight of regional and scientific authority. Kissidougou's savannization was set within a process of continent-wide 'desertification'.

Particular topics captured research attention at different times. A recurring focus was the effect of bush fire: while there was strong West African debate about this issue (e.g. Jeffreys, 1950), observers in Kissidougou were convinced that increased bush fire was responsible for the increase in savannization which they perceived (e.g. Chevalier, 1928). The issue of soil degradation dominated from the late 1940s and 1950s, undoubtedly influenced by the American Dust Bowl experience and the Africa-wide discussions it stimulated, including the 1948 Goma inter-African soil conference (cf. Anderson, 1984). Rouanet, Guinée's

---

[7] Chevalier, A., 'Rapport sur les nouvelles recherches sur les plantes à caoutchouc de la Guinée française', 1909, Archives Nationales de Sénégal, 1G276.

[8] Nicolas, 'Etat de cultures indigènes, août 1914', Archives Nationales de Conakry, Guinée, 1R12.

[9] Chevalier, for example, was well-acquainted with the Nigerian foresters Thompson and Unwin, who themselves had experience in India where these issues had been on the agenda for much longer (Thompson, 1910; 1911).

[10] Sudres, 'Quelques notes sur la région forestière de la Guinée Française', c. 1935, Archives, Institut de Recherche Agronomique Tropicale, Paris.

national forestry director at the time, attended this conference and translated many of its concerns directly into Guinée's forestry policy documents.[11] For a period, all existing environmental policies came to be justified in terms of soil conservation, and from 1949, each annual report of the national agricultural service had a section devoted to it. In 1952 the regional agricultural director described the relationship between deforestation, fire, soil erosion and laterization, as he saw it, as the most pressing agricultural development problem: 'Guinée Forestière is characterised by an originally rich soil on the way to sterilisation following the irrational agricultural procedures of the inhabitants.'[12]

The vision of a degraded and degrading relic landscape, which was created, reproduced and elaborated during the colonial period, was carried over wholesale into the post-independence era. Indeed, despite the huge changes in Guinée's political, social and economic life which occurred during the First Republic under President Sékou Touré's isolationist, state socialist regime (1958–84), the analysis informing environmental policy shows remarkable continuity. This continuity was partly institutional. The renowned French botanists who had created the degrading vision had, by the end of the colonial period, become the most senior figures in the French colonial environmental administration. The legacy of their intellectual standpoints was inscribed within the very organization and operation of the Eaux et Forêts institution. The images of environmental change derived from these 'scientific' analyses had also been incorporated into the colonial education system at primary and secondary level, and thus entered the popular consciousness of the new independent Guinéen state functionaries. The First Republic's own educational institutions – new university structures and 'Centres d'Education Révolutionnaire' – only served to reinforce these images, while the colonial image of 'traditional' local land use as destructive was reinforced in the context of the regime's rural modernization emphasis, epitomized in the push for 'tractorization'.

Since Sékou Touré's death in 1984, environmental ideas in Kissidougou have acquired a further layer of reinforcement from the dissemination of global and regional images of environmental crisis. Given FAO figures concerning rapid forest loss in West Africa (FAO, 1993), for example, it now appears inconceivable that Kissidougou should be experiencing anything else. Such figures, so frequently publicized in the more glossy development literature and on the radio, are far more accessible to the environmental administrations and urban public than are analyses of the locality itself. Equally, the rhetoric of shared environmental crisis, made so apparent in the 1992 UNCED

[11] Rouanet, R. 1951, 'Le problème de la conservation des sols en Guinée', Conakry, Service des Eaux et Forêts.

[12] Cole, H. 1952, 'La vulgarisation agricole en Guinée Forestière', Archives, Institut de Recherche Agronomique Tropicale, Paris.

conference in Rio, appeals far more powerfully to local officials than the statements of the villagers supposedly experiencing these problems. Thus in 1993 during the 'Environment Days', a local conference designed to raise awareness of Kissidougou's environmental problems, most participants framed their speeches in terms of global concern with biodiversity loss and the common West African struggle against desertification.

Colonial science developed not only ideas concerning forest loss, but also methodologies for elucidating vegetation change which became and have remained 'authoritative'. Central has been the deduction of long-term change from snapshot or short-term observations, inferring process from form. Thus it is that forest islands appear as relics indicating a historical process of forest loss; a deduction now made not only from on-the-ground botanical, forestry and vegetation survey observations, but also when forest islands appear in remotely-sensed imagery. Other aspects of vegetation form are also used to indicate forest retreat. Valentin's early deduction from the presence of oil palms in savannas has been repeated by botanists and agronomists in the 1950s (Keay, 1959)[13] and by bush fire specialists in the 1990s (Green, 1991). Botanists have surveyed the vegetation (phytosociology) of forest patch boundaries, and where they showed a mixture of forest and savanna species, deduced this 'transition woodland' to represent the advance of savanna into forest (e.g. Adam, 1948; 1968; Thies, 1993).[14] Alternative possible explanations of each of these indicators – as the outcome of local agro-ecological management strategies – have been ignored (Fairhead and Leach, 1996). Similarly, observations of processes seen in the short term are taken to indicate long-term trends. Thus modern hydrologists, like Nicolas earlier in 1914, have inferred that short drought periods result from long-term vegetation loss.

Certain scientific theories also contribute to the continual reproduction of savannization views in Kissidougou. In particular, reasoning depends on the notion of there being a climatic climax vegetation; an ultimate stage of succession which represents the region's 'natural' vegetation, and against which the degree of degradation of today's vegetation can be assessed. In Kissidougou, today's climate (e.g. annual rainfall levels in excess of 1600 mm), and the presence of humid forest species and patches are taken as indicative of high forest potential and hence of its past existence.

Scientists and others have also repeatedly observed Kissidougou's landscape from a social position which made forest destruction logical, and attention to local inhabitants' opinions difficult or unimportant. Racialist, pejorative views of African farming and forestry practices

---

[13] Circonscription Agricole Région Forestière, 'Rapport Annuel 1953', Archives, Centre de Recherche Agronomique de Seredou, Guinée.
[14] Full sets of 'indicators' of past vegetation are examined in Fairhead and Leach (1996).

came to dominate Guinée's colonial administrations. The preconceived opinions and hurried visits of today's foreign experts, and the attitudes and training of urban-based state functionaries, compound such views. It can be argued that the image of the rural farmer as environmental destroyer, and hence the need for modernization of resource management and farming techniques, conforms to and helps to justify the self-distinction of urban intellectuals as 'modern' and progressive; distinctions reinforced under the First Republic when the urbanized were politically and economically privileged, and their vision of a highly mechanized, capital-intensive technical future dominated approaches to rural development (Rivière, 1971).

The institutional and financial structures of the forestry administration are also clearly implicated in the reproduction of degradation visions in Kissidougou. The policy repertoire – which was evident by the 1930s and has persisted largely unchanged since then – involved first, the restriction of upland shifting cultivation in favour of swamp rice development, with attempts to 'rationalize' what upland farming must remain. Second, restrictions placed on local use of bush fire, whether through total fire prohibition or externally set early-burning. Third, the protection of forests both through the reservation of certain forest patches and restrictions on the felling of listed tree species – primarily those forest species valuable for timber and seen as most representative of the 'original' forest cover. Implementation, as the forestry service gained strength in the 1950s, took place within an explicit policy of repression, conducted by a para-military force. During the First Republic, fire-setting actually carried the death penalty. The forestry service came to derive essential revenues from the sale of permits and licences for timber exploitation, and from fines for breaking state environmental laws. It thus became and has remained financially reliant on the degradation analysis and on the appropriation of control over natural resource management which it implied.

National financial concerns with environmental degradation are joined by more regional and global ones. In the 1950s, for example, new funding envelopes for regional soil, climate and hydrological conservation followed the Africa-wide 1948 Goma conference. Recently, administrative solvency and development activities have become even more dependent on foreign aid, and have been subjected to various forms of 'green conditionality' (Davies, 1992; Davies and Leach, 1991). This 'greening of aid', and the specific forms it takes, clearly reflect donors' needs to satisfy home political constituencies influenced by media images and NGOs, as well as their own institutional assessments of African environmental problems. In Guinée, foreign assistance is increasingly allocated, sectorally and by region, directly to environmental rehabilitation. A new generation of heavily funded environmental projects has emerged, including, in Kissidougou, two component projects of the internationally funded Niger river protection programme. In agricultural and other development activities, as well, overt environmental sustainability components are important for attracting future funds. During Kissidougou's 'Environment Days' the prefecture's second administrator stated explicitly that: 'Donors are interested principally in environmental projects, so we must solicit their aid to ensure the development of the prefecture.' He suggested that other localities learn from the example of the Niger protection project zones, where schools, water and other infrastructural developments were provided in exchange for local participation in environmental protection. In short, presenting a degrading or threatened environment has become an imperative to gain access to donors' funds.

From such widely shared intellectual or social standpoints, reading the landscape in terms of degradation is obvious, and each observation or casual reading can serve as confirmatory evidence. Thus dry season bush-fire is easily taken as proof of a worsening problem, and the conversion to farmland of a few forest islands near the town can easily suggest forest island diminution everywhere. Notably, such casual landscape readings are often made during the dry season, when external consultants, forestry agents and urban nationals' visits to villages are concentrated. This is the destructive part of villagers' normal seasonal cycle, when bush is cleared for farming, fires sweep the savanna and trees are cut for construction or sale. Regeneration during the rainy season, anyway more subtle to observe, escapes attention within this seasonal bias (cf. Chambers, 1983).

The failure of environmental policies can also reinforce convictions of degradation. Villagers commonly continue to set 'forbidden' fires or to fell protected trees in their fallows, for instance; reflecting their need to continue agro-ecological practices in the face of – and sometimes explicitly resisting – policies which they consider ecologically inappropriate as well as socially repressive. But agencies have generally taken such 'failures' as further evidence of local ignorance and wanton destructiveness, and hence the need to implement policy with greater force, without questioning the appropriateness of the policies or their underlying analysis. It is in this way that the environmental service has served more than any other to extend state bureaucracy and military control into rural areas. Since colonial times, villagers have experienced 'gardes forêts' as the most pernicious and intrusive arm of the State. During both pre-independence and recent multi-party elections, the promise to alleviate this burden has provided a vote-catching agenda for opposition parties.

Perhaps the most surprising feature of the external narratives about Kissidougou's environment is their persistence for a century despite the highly contrasting opinions of much of the prefecture's rural population. The interface between villagers and external agencies, rooted in the colonial encounter and developed over antagonistic circumstances, has rendered the expression of local environmental experiences highly problematic. Faced by direct questions about deforestation couched in the environmental services' terms, villagers tend to confirm outsiders' opinions which are, by now, long familiar to them. Agreeing with (or not denying) visitors' views can be a polite way of coping with extractive or repressive encounters, or to maintain good relations with authoritative outsiders who may bring as yet unknown benefits; a school, road or advantageous recognition to the village, for example. Like the prefecture administration, many village authorities realize the benefits which can accompany community participation in environmental rehabilitation, and in this context may publicly agree to the 'urgent need' to plant trees, establish village environmental management committees and so on. In this respect, deforestation images have been incorporated into local political discourse concerning village–state–donor relations.

While these factors condition communication at the interface, the lack of challenge should not be reduced to this. Even where forest agents on the ground might acquire some clarity in considering local environmental perspectives, and come to doubt their messages and the fines which they impose, they have to date had almost no capacity to influence their institutions' environmental agendas. How can they be confident enough in what villagers say to question both their institutions' logic and the scientific canon as they understand it? Beyond what villagers say and what they have seen, what is the evidence for challenge? Who would listen to them? How would expressing this square with their training documents and job instructions? What effect would it have anyway at a higher level? How would adopting this perspective institutionally alter their future employment and status?

Prior to the 1950s, and within the context of a racialist colonial ministry, questioning French seniors' orthodoxy was certainly not the job of local-level Guinean administrators. When in the 1950s, the environmental services became a para-military operation, its structures became even less tolerant of critique from within the ranks. Post independence, formal science was the tool of Sékou Touré's revolution; again, this was no context for any individual's populism, least of all given the regime's surveillance and repression of potentially critical individual expression (e.g. Bah, 1990). Since 1984, and especially in the 1990s, attempts are being made to render the environmental services more responsive to local knowledge and priorities in acknowledgement of some of these problems (cf. Bah, 1989). But the shifts in organizational structures, approach and culture argued elsewhere as central for agencies to engage effectively with farmers' perspectives (cf. Chambers *et al.*, 1989; Scoones and Thompson, 1994), while in some respects outlined and planned in Guinée, have yet to be effected. Meanwhile, exceptional individuals remain frustrated. One dedicated Kissidougou forestry worker who did become interested in farmers' opinions, and who sought to work and experiment with farmers' manipulation of soils and fallows, found little space or encouragement to pursue this from higher levels of his organization.

In this context, there are also questions to be asked about the relevance to environment and development services of a more accurate understanding of environmental change. Given present institutional culture and the terms and conditions of their work, it is not surprising that there is a certain despondency and confusion of objectives among State service employees. Perhaps, as villagers frequently suggest, forest agents are not really interested in the environment: 'they only need money, and have no concern for the protection of forests', as one said (cited in Millimouno, 1993). And perhaps not surprisingly, it is the material benefits of service or project employment which dominate employees' work objectives, and there is fear of jeopardizing these by questioning institutional aims and strategy.

Images of forest loss can also be reproduced in local political discourse concerning ethnicity. Colonial stereotypes of ethnic differentiation among Kissidougou's populations rested partly on environmental behaviour. As a 'forest people', Kissia who live in the south of the prefecture were contrasted ethnically with the more northerly 'savanna people' of Kuranko or Maninka origin. The latter's historical and ongoing southwards migration, and their predilection for fire-setting in savanna farming, honey collecting, and hunting, was considered responsible for southwards savannization (e.g. Adam, 1948). Such stereotypes overlook evident similarities in everyday ecological knowledge and resource management, as well as the more complicated nature of Kissidougou's settlement history. Nevertheless, these stereotypes and their linked vision of progressive forest loss are sometimes invoked by rural and urban Kissia themselves, when expressing anxiety about Maninka domination, whether economic, cultural or military. Sharing one forest – where the forest islands of neighbouring villages have come to touch each other – is one of the strongest metaphors of Kissi political solidarity, linked as it used to be to past alliance in warfare and to common initiation institutions. Accepting the idea that the Kissi region could (until even recently) have been united in one forest provides a politically appealing vision of ethnic unity; an instance of people using environmental issues to make politico-ethnic points.

Thus intellectual, social, political and financial structures have all played a part in creating and sustaining the vision of environmental degradation in Kissidougou. Knowledge about and convictions of deforestation have been produced, not only through particular methodologies but also within diffuse relations of political and economic power. While technicists might see the problem lying in 'bad science', and its solution in 'good science' and training, there seem to be much broader and more intransigent reasons why the degradation view makes sense, which impinge on – or condition – any scientific endeavour.

Furthermore, the degradation conviction has acquired an impregnable, totalizing capacity through convergences in the views expressed by different actors, albeit for different reasons, in the contexts that matter to it. Images of environmental change invoked in ethnic discourse, for example, have converged with those in discussions of rural modernization or of financing the prefecture's development. The received wisdom today thus cannot be attributed only to scientists, donor agencies and their narratives. It is partly the product of a long history of interaction with and incorporation into local social and political processes in which villagers' very different, everyday ecological reasoning is subjugated.

Towards a re-framing of forest-savanna mosaic theory and policy

While the discursive processes sustaining visions of forest loss in Kissidougou cannot be reduced to the conceptual frameworks of ecological science, they have nevertheless rested heavily on them. It can be argued that the most fundamental props to received wisdom concerning the forest-savanna mosaic stem from the imperative to explain the disappearance of a 'natural' vegetation, imaged as a climatic climax forest at equilibrium. This allowed all vegetation change to be imaged as a lineal divergence from an undisturbed, original form, and people's impact as creating anthropogenic sub-climaxes, so hiding local enrichment practices and their significance.

Recent strands of ecological thinking, however, strongly critique such notions of an original, natural climax vegetation. In so doing, we would argue, they enable a re-placement of people's role in vegetation change. There is increasing recognition that natural disturbances (e.g. by fire, wind, animals) are a feature of most environments, forest or otherwise, which keep the ecosystem from ever reaching a stable equilibrium in any place. Instead, ecosystems come to be imaged as a patchwork or mosaic of ages of recovery from localized disturbances (cf. Whitmore, 1990). Even at a broader landscape scale, equilibrium may not be reached: it may be interrupted by significant climatic fluctuation such as that now recognized to characterize West Africa's climate history at a 100–1,000 year timescale, or by large one-off disturbances such as drought years and fire which may have long-lasting impacts on land or species composition (Sprugel, 1991).

These potential instabilities, and appreciation that the legacy of one vegetation state will influence subsequent ones, means that pathways of vegetation development can be unique and chancy. The cession of savanna to forest thicket vegetation in southern Kissidougou, for example, may partly reflect lag effects of the climatic rehumidification between the mid-nineteenth and twentieth centuries. Equally it may reflect the increase in cattle numbers in the 1950s, and their sudden loss at a time when fire control became more effective. Its character may reflect the greater availability of forest species as a result of reduced fire in regions further south. Indeed the path of vegetation development may result from a unique sequential blend of these factors. Seeing vegetation change in terms of such historically contingent transitions makes it inappropriate to evaluate vegetation in terms of the 'climactic potential' at any particular moment, and to assume that because a particular vegetation form (e.g. forest) *could* exist under present environmental conditions, that it *did* exist.

Recent ecological theory also suggests that pathways of vegetation development might best be considered as transitions between particular stabilized vegetation states, each determined by a multi-factor complex. In this, the impact of change in any particular factor will be less a smooth trend than the inducement of a shift from one state to another. Should the transition-causing factor revert to its pre-transition level, the vegetation need not return to its initial state and may move to a third (Sprugel, 1991; cf. Dublin *et al.*, 1990; Holling, 1973). Shifts between vegetation states can thus be 'chaotic' in nature. They are likely to respond to particular, possibly unique, historical conjunctures of ecological factors. Forest and savanna forms can – from an ecological viewpoint – be seen as such multiply determined and strongly differentiated 'states'. Where, as in the transition zone, climatic conditions are relatively marginal for forest, the presence of forest forms might depend on a constellation of factors including fertility cycling, soil structure, water-holding and mycorrhiza, micro-climate, the absence of fire, and germination potential interacting with the water made available by climate. In any place, a shift from a forest to savanna form or vice versa is likely to be relatively enduring both because of internal stability effects (Moss, 1982) – for example the tendency for savanna to perpetuate itself by fuelling fire, or forest to persist as it suppresses fire – and because of the enduring legacy of any transition for the structure, fauna and texture of soil and hence its edaphic qualities.

Forest and savanna vegetation forms offer very different production and gathering possibilities, and there are frequently major agricultural productivity gains to be had in deflecting savanna regeneration to forest vegetation regeneration in fallows, as well as advantages in having forest around one's settlement. Where conditions are marginal for forest, leaving a precarious balance between forest vegetation regeneration and pyrogenic savanna, the Kissidougou case suggests that people's manipulation of these processes can tip the balance. In effect, by altering the balance of interacting factors, people can initiate a shift between vegetation states which might otherwise be unattainable – or much less likely – through sequential transitions involving only 'natural' ecological processes. And one transition having been precipitated, it provides, (in time perhaps, and as other ecological factors respond to it) the springboard for others. In the Kissidougou case, the deflection of vegetation successions from savanna to forest is an essential element of local agro-ecological practice, and the way it has been effected amidst changing social, economic and demographic conditions helps to explain demonstrable patterns of vegetation change. This attention to history and local experience lay the ground for a fundamental reinterpretation of forest patches in the landscape; neither as relics nor as naturally determined by soil conditions, but as the historically conditioned products of management in relation to regionally specific ecology.

This reinterpretation, in turn, suggests ways of revising policy which would be more appropriate to the region's ecological and social conditions. The many techniques and land-use practices that have served Kissidougou's farmers to enrich their landscape and increase its forest cover are surely an effective basis for external support. In working with the local ecology of fire, soils, vegetation successions and animal dynamics, these are more locally appropriate, integrated with the social matrix and thus more cost-effective in terms of labour than are the forestry packages generally proposed by outside agencies. Given that farming in the region is not inevitably degrading, environmental policy may look to support as well as to rationalize and regulate it, and specifically to support those upland farming practices which improve soils and fallow vegetation rather than concentrate technical effort exclusively on swamps. Implied, too, is the need for environmental agencies to shift away from direction (through repression or organizational restructuring) towards recognizing and supporting the diverse local institutions engaged in resource management, and a more responsive role in providing requested services. The history of people's 'opportunistic' responses and uses of non-equilibrium ecology suggests grounds for respecting and supporting this through a policy framework which enhances, rather than reduces, people's resource management control and flexibility.

As an alternative to climax reasoning, then, non-equilibrium ecology opens up possibilities for interpreting forest-savanna ecology within its real historical specificity: indeed for a view of ecology as history. By removing the strictures of a concept of natural climax vegetation, it also opens up scope for better considering people's impact on vegetation. While we have not 'tested' this theory in ecological terms – indeed, in keeping with arguments concerning ecological pluralism (McIntosh, 1987), would not seek to establish it as any new orthodoxy – we can argue that this perspective provides a valuable counterpoint to climax views. In Kissidougou, it provides a re-framing of the forest-savanna mosaic which potentially rescues inhabitants' experiences and opinions from the received wisdoms which have silenced them.

7

Dryland Forestry

Manufacturing Forests & Farming Trees in Nigeria[1]

REGINALD CLINE-COLE

Introduction

Although forestry contributed only two per cent of Nigeria's GDP during the 1980s, and has not supplied unprocessed logs for export since 1976, it makes indispensable direct and indirect contributions to subsistence, exchange and trade in Nigeria's northern drylands. It provides a wide range of products and services: from timber, fuel and food supply; through the provision of aesthetic satisfaction; to environmental protection (Areola, 1987; Morgan, 1985; Cline-Cole, 1994). Currently, forestry provides employment for up to two million Nigerians, most of whom are involved part-time in fuelwood and polewood collection, transport and sale (Hyman, 1993). It is unlikely that any of the estimated 75,000 Nigerians employed full-time in log processing (Silviconsult, 1991) are dryland inhabitants.

Two broad, partially overlapping and, at times, conflicting management systems may be distinguished that have co-existed within Nigerian dryland forestry since the early years of this century. Since introduced, around 1902, outsider or expatriate methods and techniques of forestry administration and management have been increasingly, often incoherently, superimposed on indigenous, insider or local management structures, practices and beliefs (Cline-Cole, 1994; Morgan, 1985; Egboh, 1985).[2]

This chapter focuses on the received wisdom which informs these systems, paying particular attention to their divergent justifications and goals, and their management philosophies and styles. It also addresses

---

[1] I owe the idea for this title to Murray Last, and would also like to thank the Director and Compensation Officer of Bayero University's Physical Planning Unit, and Alhaji Hamza Turabu of the Kano State Forestry Division, for permission to consult official records containing the information used in the Kano case study discussed in this paper.

[2] In reality, of course, this dichotomy simplifies a complex situation. Not only is each 'forestry type' itself made up of several sub-types, selective mutual borrowing between and within (sub-)systems means that it is not always possible to disentangle the influence of one type of forestry from the other.

corollary arguments about the origins and persistence of both the outsider approach to dryland forestry, and the complex rationale of insider forestry management. The chapter highlights the conflicting perceptions of the nature, causes, extent and consequences of dryland forest resource transformation (and, by implication, wider 'environmental' change), and the varied and dynamic interests they serve; and calls into question the conventional view that expatriate intervention in, and transformation of, insider forestry is necessary for sustained yield management, and in the best interests of dryland environment and society.

The regional context of dryland forestry

Figure 7.1 maps the location and extent of Nigeria's drylands, whose 28 million people remain 75 per cent rural and agrarian. It also summarizes background information on regional population density, vegetation and rainfall distribution, and farming intensity, in an area in which long-term increases in population totals and densities have led to rising, albeit variable, pressure on agricultural land. Agricultural land doubled in extent between 1965 and 1990, largely at the expense of woodland, pasture and fallow (Bdliya, 1991; Mortimore *et al.*, 1990; Silviconsult, 1991).

The environment for crop cultivation, livestock and forestry systems has been fundamentally altered since the 1970s by diminishing rainfall totals, altered monthly rainfall distribution, and persistent drought (Mortimore, 1989). Nonetheless, regional agricultural and forestry output continues to be made up of livestock products; annual and minor root and vegetable crops, and their residues; and tree products (with fuelwood being the most important by volume and value) from cultivated, fallow and pasture land (Adams and Kimmage, 1992; Mortimore, 1972; Pullan, 1974).

Natural woodlands are most widespread in areas of low cultivation intensity, with ecozones 2 and 4 being the most heavily wooded (see Figure 7.1). However, as cultivation occupies three times as much land as woodland, more than a third of regional standing wood volume is found on farmland, including farms in areas of intense cultivation. Farm tree densities range from 1–2 per hectare in the dry north-east to 12–22 per hectare in parts of the intensively cultivated zones around towns like Kano.

According to Silviconsult (1991), land-use conversions to agriculture and uncontrolled grazing are the main causes of deforestation in the drylands, while problems of fuelwood supply are also held to be both a cause and an effect of wood resource depletion. Sustainable regional fuel production is said to be exceeded by present-day demand, with imports from wetter and more heavily wooded areas to the south

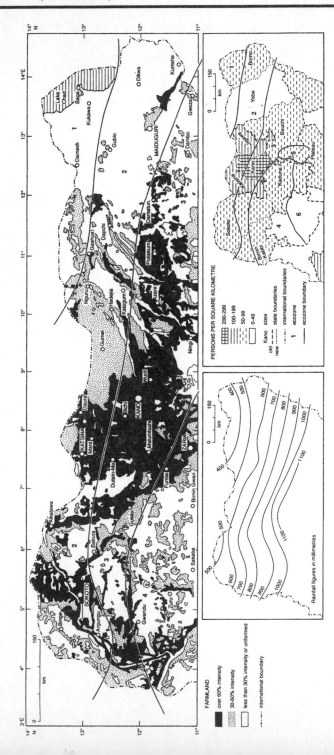

Figure 7.1   Farming intensity, population density, rainfall and ecozones in northern Nigeria
Source: Cline-Cole (1995).

making up some of the deficit. Deforestation is proceeding most rapidly in 'frontier' zones of low but increasing cultivation intensity, while farm tree numbers appear to be most stable, or even increasing, in parts of the heavily populated, and most intensively cultivated close-settled zones (e.g. around Kano). In the latter areas, off-farm activities, including the collection, processing and exchange of tree products, make a vital contribution to rural livelihoods, and indigenous farm forestry has a long history.

Although claims of widespread desertification abound (Anderson, 1987; Anon, 1990; KNARDA, undated; Sagua *et al.*, 1987), Mortimore (1989) warns that evidence linking bush-burning, overgrazing, wood-cutting and agricultural clearance to dryland degradation needs to be treated with extreme caution. Such degradation is difficult to assess, he argues, and available 'evidence' for desertification tends to rest on an inadequate empirical base (cf. Swift, this volume). Nonetheless, introduced forestry has, historically, failed to demonstrate such caution. It has consistently blamed indigenous agriculture and forestry for perceived problems of dryland degradation, and steadfastly promoted outsider forestry practices as an effective response to such problems. In turn, indigenous foresters have long accused outsider forestry of a lack of appreciation of their goals and methods, and, consequently, of a failure to respond to their felt needs.

'Cultural dissonance' of this kind has been of remarkable longevity, sustained by contemporary social struggle both within and between dryland forestry systems. It is ultimately born of conflicting values of dryland forestry potential and use; of what a regional forestry landscape should look like; and of what the functions of regional forestry actually are, or ought to be.

Cultural dissonance in dryland forestry

Travel accounts, archival sources and oral history document the existence of a sophisticated pre-colonial forestry culture in the dry-lands, some of whose more noticeable practices, such as the planting and protection of farm trees, and their selective and rotational exploitation, have survived both colonial and post-independence interference. Recent work shows how pre-colonial common-property and private institutions regulated access to, control over, and exploita-tion of resources within indigenous forestry (Cline-Cole, 1994). Such regulations were enforced by supervisory officials with sometimes overlapping responsibility for land, water and forest or woodland resources. These regulations were flexible, locally based and socially sensitive, but, for all that, well-defined. Natural-resource use and management were, significantly, holistic activities which did not separ-ate household from group livelihood strategies, nor the management of production from the management of the environment (cf. Redclift, 1993). Above all, pre-colonial natural-resource decision-making rested firmly with local communities, albeit within the wider context of a society which was both stratified and inequitable, and within which indigenous forestry knowledge was not necessarily common knowledge, either across agro-ecological zones or within and between populations and communities.

From the earliest days of its introduction outsider forestry had its own ideas about ways in which forest and other natural resources could be more effectively managed, and made more beneficial for people and the environment even in this area of limited export potential. In the words of Sir Frederick (later Lord) Lugard, Northern Nigeria's first High Commissioner and the architect of indirect rule, introduced forestry was '[i]n the truest and best interests of the country, [and aimed] to protect an ignorant people against themselves, to secure the welfare of future generations, and to safeguard the interests of one community against the selfish action of another' (1970: 434). It was therefore comprehensive in its ambitions to control forest resources if not land (cf. Grove, 1994); and, so Lugard's detractors suspected, fitted in with both his personal expansionist ambitions and overwhelming desire to ensure that 'his' colony of Northern Nigeria kept up with socio-economic and political developments taking place in Southern Nigeria (Egboh, 1985).

The period 1914–16 marked a major watershed in the history of introduced dryland forestry. It saw responsibility for decision-making shift away from local communities to a Northern Region forestry service whose headquarters were opened in Zaria in 1914. A forestry ordinance was instituted which restricted, and in some cases completely transformed, the rules governing access to and control over communal and private forest resources. A forestry school was established for training Nigerian staff in introduced forestry principles and methods, which were, by implication, superior to, or at the very least, preferable to indigenous ideas. Arguably most important, the period saw the creation of separate regional departments for agriculture and forestry, which 'gave formal expression to the previously alien notion of forestry as both administratively independent of and conceptually distinct from regional agriculture' (Cline-Cole, 1994: 74; cf. Davis, 1982; Egboh, 1985).

For Lugard, the introduction of outsider forestry, and its carefully contrived hegemony over indigenous forestry, was justifiable on environmental and social grounds, if not on a strictly commercial basis. There may have been only a few tree products of direct export value but the fact that local subsistence and livelihoods were thought to be threatened because the environment was being degraded was, in Lugard's opinion, more than adequate cause for intervention. Thus he cited a perceived imperative for, first, the establishment of tree belts to stabilize 'moving dunes' in northern Sokoto and Borno 'where arid conditions were considered inimical to forest vegetation' (Lugard, 1970:

431). Second, he advocated the strict regulation of 'wholesale' vegetation destruction by firewood collectors, farmers and pastoralists in heavily populated rural areas, and around new colonial administrative centres and large 'native' cities. Third, he recommended the establishment of:

> . . . timber plantations [whose] object [wa]s to introduce good exotics . . ., to test the suitability of the climate for different kinds of trees, to maintain a steady supply of seedlings, and . . . to concentrate good timber at places where it w[ould] be of special value and c[ould] be intensively exploited. (Lugard, 1970: 437)

There was also need, he argued, for controlling, maybe even banning, annual bush fires, which were started by pastoralists and hunters to flush out game and stimulate vegetation regrowth for livestock.

Farms, forest or woodland, and pasture thus had to be kept largely separate, Lugard concluded, and access to forest resources had to be restricted, not only on state-controlled land, but also on communal and private land. Such a task, he was convinced, 'c[ould] only be carried out by the combined effort of the Central Government and of enlightened Native Administrations' (*ibid.*: 432). Lugard's interventionist ideas (or those of Mr H. N. Thompson who, as Director of Forests, had supplied the memorandum used by Lugard) dovetailed neatly with prevalent views of West African environments. These included the influential conservationist views expressed since the 1880s by Alfred Moloney (see Grove, 1994, and Fairhead and Leach, this volume) and, subsequently and repeatedly, the received version of desertification shared by national governments, aid agencies and some natural scientists which, like Moloney's views, justified intervention in rural land use (Swift, this volume; Mortimore, 1989; NEST, 1991; Stebbing, 1935). The spectre of desertification remains to this day a recurrent metaphor for a perceived need for greater intervention in dryland forestry (Sagua *et al.*, 1987).

Thus was born an introduced forestry orthodoxy which was restrictive and exclusionary and, on occasion, repressive. A popular protest song of the late-colonial era, for example, laments the policy of forest reservation, and condemns both the over-zealous policing of the forest estate and the rigid enforcement of restrictions on farm-tree exploitation (Watts, 1983). The tradition of exclusion and restriction lives on, kept alive by a siege mentality shared by long-serving forest administrators. These administrators tend steadfastly to oppose the devolution of responsibility for management of the forest estate to local communities, and to a younger generation of foresters who would supervise such a project.

This 'establishment' perspective argues that the institution of such management partnerships would directly threaten the forestry administration's survival under prevailing conditions. Notably, these are said to include a highly fragmented and poorly consolidated forest estate; a chronic and worsening underfunding of centralized or 'organized' forestry; the threat posed by the expansionism of farming and livestock-raising; and the much greater priority accorded agriculture within Federal and regional development policy and practice (Silviconsult, 1991). As one of the longest-serving forest administrators confided to me:

> Forestry remains about power, control and survival; the only people within the profession who will argue to the contrary are misguided young men [*sic*] who, I am sorry to say, have *no* idea about the significance of politics for the continued existence of forestry, both as an autonomous player in environmental management and rural development, and as an independent employer.

Proponents of the 'new' forestry consistently exaggerated what were perceived to be the limitations of indigenous forestry (and other forms of natural-resource management), and at the same time devalued its more desirable aspects. For example, a Nigerian representative of outsider forestry noted approvingly in 1944 that, in exploiting farm trees for fuelwood, 'most [Kano] farmers practise[d] a kind of rotation as we do in our well-organised plantations [and] are careful not to cut all their standing trees at once' (Adelodun, 1944: 30). Nonetheless, he considered it both unsatisfactory and unacceptable that farm trees rather than woodland vegetation, and farmland rather than plantations, provided the bulk of Kano's fuelwood needs. Indigenous farm foresters just did not know that 'fuelwood should, if conditions were normal, come from forests and not farms' (*ibid.*), and that where the 'land for miles around [is] taken up by farming [the supply of fuelwood] presents a problem [whose] solution . . . lie[s] in raising plantations. . .'. (Kerr, 1940: 21).

Thirty years later another commentator observed that:

> . . . in the northern sandy areas around Gumel and Hadejia, where most trees have already been removed, wind-blown sand at the beginning of the rainy season often buries or cuts off the small shoots of newly germinated corn. As firewood becomes more scarce the people burn dung for cooking and heating instead of using it as manure. This is the practice now in northern parts of Sokoto. (Howard, 1976: 21)

And more recently, we read that:

> . . . [t]he rate of woodfuel consumption far exceeds the replenishing rate to such an extent that desert encroachment, which is the result of massive and wanton destruction of trees and vegetation, is now a very well-recognised problem posing serious threats to the lives and well-being of crops and forests as well as animals and man himself . . . (KSCASE, 1989: 1)

Although largely assumed rather than proven (cf. Cline-Cole *et al.*, 1990; Mortimore, 1989), such observations have led to a seemingly inevitable conclusion, explicitly spelt out in both the National Conservation Strategy and the 1988 Forestry Policy Statement (Hyman, 1993). This is the need for greater government and donor investment in forestry: in the consolidation and expansion of the national forest estate, and in more effective ways of securing the participation of dryland inhabitants and private investors in introduced forestry activities. While such measures might satisfy global and regional environmental agendas and attract donor funding, the concerns which they reflect are not necessarily shared by local inhabitants (cf. Amanor, 1994a; Fairhead and Leach, 1994; Davies and Richards, 1991).

But differences between introduced and indigenous dryland forestry have been as much about perceptions and institutional politics as about substance and direction (Egboh, 1979; Fairburn, 1937; Morgan, 1985; Pullan, 1974). Environmental protection and fuelwood production, the two pillars of introduced forestry, rate much less highly as goals in indigenous forestry. The latter places a much greater premium on fruit and food production and shade provision, with environmental protection and fuelwood production being considered of only subsidiary importance. Similarly, although introduced forestry favours a few 'dedicated' or single-use exotic species, local inhabitants perceive that, overall, with the exception of the (naturalized) exotics neem (for shade) and mango (for fruit), indigenous multi-purpose species are much better suited to the dominant goals of insider forestry. Such species make up 18 of the 20 species reported as the most frequently occurring farm trees in the drylands (Silviconsult, 1991). And, while the Kano State forestry establishment, for instance, speaks of 'a total lack of [seedling and tree] care and protection . . . and [a] general ignorance of the importance of the uses of trees on the part of collectors' (KSAC, 1987: 6), there is increasing evidence that forestry management practices like the protection of naturally regenerating seedlings, transplanting of wildings and farm-tree planting continue to be widespread in the drylands (Lockwood, 1991; Silviconsult, 1991).

Although relations between the two forestry systems have been uneasy, they have rarely degenerated into open conflict and violent protest. Grove (1994) has observed that resistance to colonial conservation activity in West Africa 'was manifested less in terms of direct clashes with the colonial state and more in terms of conflicts between indigenous groups, tribes and classes and, not least, between men and women'. In Northern Nigeria conflicts commonly pitted state forest assistants, rangers and guards against dryland inhabitants whose livelihoods were threatened by forest reservation and restrictions on the cutting of farm trees. But even conflicts such as these were more muted in the drylands than in Southern Nigeria (Buchanan and Pugh, 1955; Egboh, 1979; 1985). As archival sources reveal, more evident were dignified exchanges of correspondence between native rulers and their colonial overlords, regarding the income-redistribution effects of forestry intervention.

In the event, protest at state forestry intervention in the drylands has, historically, taken the form of silent resistance. This reflects the fact that indigenous foresters and others have often been able to ignore or pervert outsider forestry rules and regulations; physically evade detection and control; and watch the state respond with that paradoxical mix of *laissez faire* and force remarked on in other conditions of conflict and change in rural Africa (cf. Adams, 1988; Crummey, 1986; Mortimore *et al.*, 1990; Watts, 1983).

Several factors help explain the lack of open conflict between introduced forestry and dryland inhabitants (Anderson, 1987; Howard, 1976; Hyman, 1993; Morgan, 1985; Egboh, 1979). First, with few regional products of export value to protect, develop and exploit, introduced forestry was considered of local (subsistence) rather than national (export) importance. It was, as a consequence, not only less intensively practised than in the South, but was also administered largely 'on the cheap', through local Native Authorities and Local Government Councils, albeit under supervision from regional forestry services. Second, leading members of local elites (e.g. traditional rulers like the Emir of Kano), without whose cooperation and collaboration outsider-forestry-as-indirect-rule would have been impossible, have been willing to take up issues of common concern with state forestry authorities, on behalf of their subjects and political constituents. Third, with the exception of the Sokoto area, the forest estate was, and remains, limited in extent. Plantation and tree-crop farming is becoming more prevalent but is still not sufficiently widespread to cause more than localized land shortage. Outside the close-settled zones, population pressure has only recently started to build up. Fourth, dual control of forest administration by Local Government Councils and state Forest Departments has led to a confusing overlap of authority with no one taking full responsibility for policing the estate, the effective protection of which, especially in more remote areas, has always been compromised by staff limitations and logistical difficulties. Finally, forest policy guidelines are not only inconsistent, they are also inadequately coordinated with those issued in the agriculture sector.

It needs to be emphasized, however, that this seeming disarray in the implementation of outsider forestry has not prevented the emergence and persistence of remarkable unanimity in its approach to tree planting and management. This contrasts markedly with the eclecticism of indigenous forestry in the pursuit of its own goals.

Manufacturing forests in introduced forestry

Introduced forestry has changed little since colonial days, in legal framework, policy foundations, and management strategies. The 1938 Forestry Law, which was adopted with minor amendments as the Northern Region Forestry Law, Cap. 44 of 1965/66, still provides the legal basis for both the sustained yield management of a fragmented dryland forest estate, and the imposition of restrictions on the felling of valuable timber and farm trees on unreserved land (Buchanan and Pugh, 1955; Egboh, 1985; Morgan, 1985; Mortimore *et al.*, 1990).

Similarly, policy continues to be driven by the declared aim of stabilizing and maintaining a balance between agriculture and forestry as integral parts of a regional production system, capable of sustaining livelihoods while protecting and improving the environment:

> . . . since trees and vegetation are essential for a healthy environment as they check wind and water erosion and desertification, and also since wood from trees serve as a source of energy and other domestic needs, then it is especially important that trees be protected through measures that will reduce excessive destruction and encourage replanting on a rational basis. (KSCASE, 1989: 2)

The measures adopted include protection, production and extension forestry, each of which is geared largely toward the achievement of a single goal or the production of a single output; and each of which has experienced fluctuating fortunes over the years in response to national and donor priorities, advances in research, and so on (Howard, 1976). Expatriate forestry is most intensively practised in the forest reserves, Communal Forest Areas, plantations, shelterbelts, woodlots and nurseries which make up the forest estate, and which, when adequately funded and managed according to strict technocratic principles, assume distinct 'ordered' forms in the regional landscape. Expatriate forestry's archetypal plantation or shelterbelt involves regularly spaced rows of trees in even-aged, frequently monocultural stands. These make for standardized establishment procedures (sometimes including mechanization) and management practices. They contrast markedly with the seeming 'disorder' of highly diversified indigenous agroforestry landscapes.

Protection forestry initially concentrated on ensuring the stability of climatic and water regimes on land considered unsuitable for arable cultivation by declaring them forest reserves or 'protection forests' in which grazing, burning and cultivation were strictly controlled, and the collection of forest products for local subsistence regulated. Since the prolonged drought of the 1970s, the establishment and maintenance of shelterbelts and windbreaks in the northernmost regions of the drylands were considered to be of prime importance (Hyman, 1993; Sagua *et al.*, 1987).

Both forest reserves and shelterbelts can, under sustained yield rotation, produce supplies of timber, poles, fuel and other forest products for subsistence and sale. However, plantations are considered better suited to the latter role, particularly when they are readily accessible by road and are located near towns. Production forestry is, therefore, committed to the design, establishment and management of plantations, which in addition to ensuring local supplies of a range of

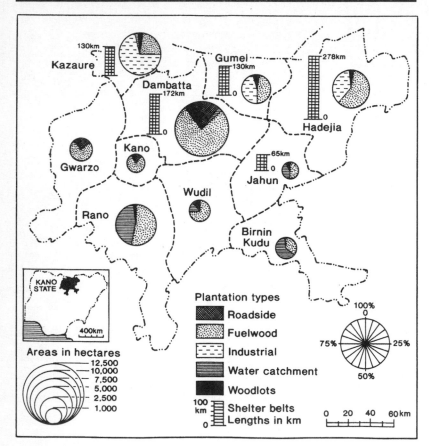

Figure 7.2   Plantations in Kano-Jigawa
Source: Based on figures supplied by the Kano State Government

forest products, are also seen as possessing potential for environmental protection (Figure 7.2). Some reserves near towns have been converted to plantations, which are thought to grow at a faster rate than natural woodland, and whose 'tree crops' are seen as being capable of 'earn[ing] as much and often more than most other crops, on an annual basis' (Horsman, 1975: 74). A moribund medium-to-long-term programme for plantation development approved in the 1970s appears to have been revived, and will be focusing on timber production in the wetter south of the region (Howard, 1976; Silviconsult, 1991).

The restriction of formal production and protection forestry to the forest estate considerably limits their impact. Expatriate forestry has sought to counter this shortcoming by integrating its protection and production activities with indigenous agriculture and grazing, which remain the dominant land uses outside the forest estate. Thus extension forestry, which encourages farmers to plant and maintain trees in a variety of configurations in areas stripped of woody vegetation, and which has received sustained encouragement since the 1970s, is currently the most favoured form of intervention (Howard, 1976; Hyman, 1993). It is a modification of the old (protection) policy of simply imposing restrictions on the exploitation of farm trees.

Unlike production and protection forestry, regional extension forestry is dominated by parastatal development agencies and NGOs, who promote the adoption of standardized management or planting 'units' of shelterbelts, windbreaks, woodlots and farm trees (Silviconsult, 1991).[3]

Overall, introduced forestry has been strongly technocratic in orientation, concentrating on the practical aspects of tree selection, propagation, planting and protection with little regard for what Bassett refers to in another context as the 'interlocking political, economic, and cultural-ecological problems affecting land use' (1993b: 132) and, more important, their potential for constraining the realization of technical objectives. Policy has consistently favoured centralization and standardization. Consequently, environmental variation is underestimated, social diversity neglected and management insufficiently adapted to specific socio-ecological milieux (cf. Fitzgerald, 1994). These tendencies of introduced forestry systematically devalue the contribution of indigenous natural-resource managers to environmental management, by excluding insider foresters from the management of a forest estate, whose components are managed as 'islands' in, but not really a part of, surrounding areas of both complementary and conflicting land uses.

As a result, introduced forestry is no nearer to achieving its declared goals now than it was in 1914. The restriction on farm-tree exploitation is now a dead letter. Large areas of erodible land requiring protection and conservation remain outside the forest estate. Within the latter, not only are tree-seedling survival rates consistently and unacceptably low, but almost all forest reserves are *de facto* grazing reserves or sources of forest products, with several showing signs of encroachment by farmers as well as evidence of land degradation.[4]

More than 90 per cent by volume of regional fuelwood supplies still come from private and communal sources outside the forest estate, some of which are more sustainably managed than plantations and reserves. Finally, despite the existence of a multiplicity of active extension forestry agencies, programmes and projects, forestry managers have, by their own admission, been only moderately successful in attracting private investment in communal and private plantations

---

[3] The Katsina Afforestation Project, for example, offers farmers a choice of tree planting models with precise (pre-determined) specifications for layout, size and composition as part of its extension programme.

[4] Some forest reserves have been converted to grazing reserves, managed in collaboration with regional livestock services.

and tree farms, and in securing 'mass participation' in tree planting, whether in woodlots, hedgerows, windbreaks, or on roadsides or farms (Anderson, 1987; Anon, 1990; Howard, 1976; KSAC, 1987).

Within the drylands as a whole, the contrast in attitudes between insider and outsider forestry is most marked in the intensively farmed close-settled zones. Here, indigenous forestry's integration with crop and livestock production is most advanced. Outsider forestry has been least successful in imposing its vision of landscape management on land managers, and indigenous beliefs and practices have proven considerably more effective and more durable than those of introduced forestry in responding to felt needs. These and other upland areas of permanent cultivation which support relatively widely spaced trees are known within outsider forestry as 'farmed parklands'. In them, successful programmes of intensive farm tree planting would lead to higher planting rates as farmers outnumber foresters. These would also result in savings in public expenditure because labour costs would be transferred to farmers and provide environmental protection for a considerably larger proportion of the drylands than that making up the forest estate (Anderson, 1987). Given its extremely limited success with both protection and production forestry to date,[5] expatriate forestry has set its sights on farmed parklands and extension forestry.

This marks a major sea-change in policy. Is a recognition of the need for '[a] change of traditional attitudes [involving the bridging of] the division between general agriculture and forestry' (KSAC, 1987: 4) belatedly replacing a dogged attachment to the establishment of fuel plantations, most of which were, in any case, more 'to provide a form of "window dressing" than to meet any real economic need for woodlands raised by artificial means' (Kerr, 1940: 21)? And, in so doing, is policy finally responding to long-held minority views within expatriate forestry, that some forms of smallholder farm tree management are consistent with sustainable exploitation and worthy of emulation (see, for example, Fairburn, 1937; Jones, 1938; Mortimore *et al.*, 1990)?

Unfortunately, the explanation is more prosaic. Regional interest in extension forestry since the 1980s coincides with current donor preference for forestry initiatives to encourage the planting and protection of farm and compound trees as part of a wider attempt 'to understand agroforestry in Africa, and its possible role in a strategy for sustainable development' (Cook and Grut, 1989: 1).[6] Early indications

---

[5] Forest reserves occupy land which was readily available rather than that which was most in need of protection, while plantation fuel and poles have been unable to compete successfully with forest and woodland products in regional markets.

[6] Funding flows also influenced the pace and direction of forest reservation during the 1930s; plantation establishment during the 1950s, 1960s and 1970s; and shelterbelt planting during the 1970s. Not all funding was external in origin. For example, Federal oil revenues underwrote an Ecological Disaster Fund, which financed protection forestry initiatives in the drylands during the 1970s and 1980s, in addition to funding a shelterbelt research station in Kano.

are that recent extension and farm forestry projects are not necessarily free from either the universalizing tendencies associated with plantation forestry design or the preference for exclusionary rather than inclusive management so beloved of protection forestry (Anderson, 1987; Anon, 1990; Cook and Grut, 1989; Silviconsult, 1991). In this form extension forestry does not represent a true departure from 'traditional' expatriate forestry policy and practice; it is little more than an attempt to extend 'scaled-down' versions of standard practice to farmland to satisfy changing donor preferences.

The universalizing approach to the 'manufacture' of man-made forests or trees dies hard. Its persistence remains proof of the continued lack, within outsider forestry, of a proper appreciation of the methods and dynamics of indigenous forestry, and of the motivations of dryland inhabitants who devote land, labour and capital to the less technocratic, more diversified and more flexible 'farming' of trees, woodland and forests. It is also a reminder that expatriate forestry is only partly responsive to local needs, which it sees as conflicting with rather than complementing its own goals, one of which is the encouragement of dryland inhabitants to 'participate' in its own project.

Farming trees in indigenous forestry

In contrast to mainstream outsider forestry, indigenous forestry favours inclusion and combination. The management of forests and woodland is not considered distinct from the management of trees on farmland (cf. Shepherd, 1992). Insider forestry is characterized by diversity,

> . . . involving the incorporation or retention of trees or shrubs into agricultural or pastoral systems. Such activities may include planting fruit trees around a homestead, growing trees in a woodlot to produce fuelwood or building poles, or intercropping trees with other crops on a farm plot. They also include passive systems that are based on protection and natural regeneration of indigenous trees. (Cook and Grut, 1989: 3)

Within this context, forest management by herders, woodland management by sedentary farmers, and the management of individual farm trees constitute three distinct forms of dryland forestry management, each encompassing a large number of individual practices, with the last of these being the most intensive (Shepherd, 1992).

The complex agrosilvipastoral landscapes which result from inhabitants' diverse choices and management decisions contrast markedly with the neat quasi-exclusionary categories of outsider forestry (Cline-Cole, 1995). And although the same combination of 'passive' (simple protection) and 'active' (planting) management strategies and practices observed in introduced forestry is readily identifiable here also, the overriding impression is one of sharp contrasts in perceptions, and conflicts in philosophies and styles, between the two systems. In particular, among insider landscape managers, a broad view of 'forestry management' prevails, which extends the ambit of 'forestry resources' beyond trees and shrubs, to include grasses, epiphytes and parasites. Institutional arrangements (e.g. land and tree tenure, gender relations, family and kinship systems) which regulate access to and control over productive resources by different groups of people are also constant preoccupations (cf. Mearns, 1995).

In effect, indigenous forest landscapes link farm and forest in time and space, closely integrating the management of individual trees and shrubs on farmland with the management of entire woodland, forest and fallow stands. They reflect a forestry culture which runs counter to the universalizing tendencies of outsider forestry. Forestry is an integral part of both crop cultivation and pastoral production. Many cultivated or recently fallowed fields, together occupying about a third of the drylands, support a variety of species of trees and shrubs at varying stages of growth and density. Trees and farmland are perceived to go together, with farm tree planting and management being considered a normal farming activity. Under these circumstances, outsider forestry's generic 'farm parkland' landscape is replaced by at least three conceptually distinct but closely linked landscape types. These reflect a degree of adaptation to ecological and social diversity alien to expatriate forestry, and act as a reminder of the need for conflicting demands (for fuel, food, shade, etc.) to be constantly renegotiated.

Indigenous forestry is characterized by enormous geographical variation. In the close-settled zones, for instance, farmers' expertise deals mainly with the intensive management of individual farm trees and shrubs. In woodland or forest areas farmers are more experienced in the management of natural vegetation communities, requiring detailed knowledge of species' site preferences, the dynamics of vegetation succession, species associations, and leafing and fruiting cycles.

Take the contiguous area of peri-urban agricultural land to the west of Metropolitan Kano (see Figure 7.1). The 'story' of this area is of a long history of human occupation, stable and well-established land occupancy, land shortage, and very small plot sizes. This is reflected in a variety of ways in the agrosilvipastoral landscape. Baobab trees bear witness to early occupation; boundary plants attest to the need to establish and protect secure tenure over land by permanently delineating it; and the careful husbandry of multi-purpose farm-tree species reflects the need both to enhance crop yield and to maximize productivity from the scarce land devoted to trees and shrubs.

Yet, in this densely populated area where farm-tree density increased by a quarter in the decade up to 1981 (Cline-Cole *et al.*, 1990), land ownership is distinctly inequitable. Female-owned plots constitute less than 1 per cent of the land area. They compare unfavourably with plots owned by men, particularly urban-based men, in several ways. Women's plots are the smallest in size, have the lowest ratios of tree value to land value; the lowest tree densities; the least diverse composition of species per plot; and a less-than-average representation of shade providers, fruit producers and farm boundary markers within overall farm-tree populations. One possible explanation for these differences lies in Islamic beliefs and codes of conduct which favour male beneficiaries in the inheritance of land and, via the practice of female seclusion, restrict women of reproductive age to an obscure role in farm forestry. Female-owned plots are actually managed by men, who treat them as part of larger agrosilvopastoral enterprises, which include their own plots, livestock and trees. It is therefore not certain that such plots are significantly different from other plots managed by men other than their owners.

However, women can retain complete control over tree products, and whole trees and shrubs, whether or not they manage such trees themselves, and whether or not these grow on land they own. Women do not need to own land to own trees (which may be scattered across a number of holdings), and do not need to undertake day-to-day tree management to influence the composition and appearance of forest landscapes.

It is thus women's indirect role in farm forestry which needs to be emphasized, particularly when, in the Kano close-settled zone as across most of the drylands, women are exempt from that assumed most archetypal of female tasks – the collection of domestic fuelwood. For instance, *lalle* (*Lawsonia alba*) and *zogale* (*Moringa oleifera*) border shrubs are women's crops, which together add up to more than the total number of individuals of all the other species put together; indeed, *lalle* is found on more than half of all farm plots. In addition, women also own individuals of other species, notably the Locust bean (*Parkia biglobosa*), which is found on about a quarter of all plots, and in numbers which compare favourably with those of the five other species most frequently interplanted with agricultural crops, and which are usually owned by men.

Spatial and structural variations in indigenous farm forestry, then, are the product of differences of interest and strategy, which reflect the existence of complex decision-making structures and processes. Forestry is contested terrain, as much within as between forestry systems. Introduced forestry has never been adequately equipped to deal with such complexity or its implications for policy.

Concluding remarks

Recent policy statements on environmental protection, natural resource conservation, forestry, and energy from a variety of sources (Federal and

State government, local and international conservation groups, and donor agency representatives) rest on claims that several million hectares of land in the arid zone of Nigeria are already desertified or in danger of becoming so; that desertification has reduced the productivity of natural resources; and that this, in turn, severely threatens rural production systems and leads to a reduction in biodiversity (NEST, 1991; Silviconsult, 1991). These views represent a particular orthodoxy, which this paper has used to explain the origins and persistence of expatriate dryland forestry.

This orthodoxy is embedded in the specific institutional context described, and continues to serve both political and economic interests, as 'mainstream' forestry administrators struggle with the task of maintaining their influence and authority within an underfunded sector of the regional and national economy in need of a massive infusion of donor capital. Not even Lugard, the father of expatriate dryland forestry, was above establishing a regional forestry service so as to bolster both his and the Northern Nigeria Administration's standing with, and their combined chances of attracting development funds from, the Colonial Office in London. At times, expatriate dryland forestry has seen its interests converge with those of non-foresters (e.g. conservationists, wildlife and parks services, aid agencies, cash-strapped regional and national governments) to their mutual benefit, over common concerns such as desertification, biodiversity, or livelihood security. While they may provide justification for donor funding of interventionist policies, they do not necessarily respond to grassroots concerns and are not always based on accurate readings of environmental behaviour and change.

Conflicting perceptions of environmental change are plural rationalities, and are bound up with contemporary social struggle. The continued appeal of falsifiable expatriate forestry perceptions of indigenous land management needs to be seen within this context, rather than simply as evidence of a predisposition for incremental rather than revolutionary change within the bureaucracies responsible for dryland forestry policy. After all, there already exist significant overlaps in the practice of forestry in the drylands, due mainly to the fact that outsider practice has selectively incorporated elements of indigenous forestry to satisfy donor requirements when necessary or expedient. Insider forestry has also adopted and indigenized elements of outsider forestry which meet its perceived needs. In the absence of adequate infrastructural and logistical support for the proper exercise of technocratic forestry, many Western-trained forest managers have often found it convenient to adopt insider methods and practices, although this is usually considered as a measure of last resort rather than the result of systematic evaluation of potentially viable alternatives.

There is, therefore, some potential for consensus which might lead to the emergence of more formal hybrid systems, which reflect commonly felt needs within dryland forestry. These include ensuring access to entire agrosilvipastoral landscapes, whose various components need to be exploited in an integrated manner; conservation of specific types of forest, woodland, trees and shrubs; a continuing ability to grow trees and shrubs; a capacity to sustain forestry production in all its diversity; and, finally, retaining a commitment to the development of resource users and their livelihoods, and facilitating their manipulation of the environment as an integral part of livelihood strategies (Cline-Cole, 1995). However, such approaches are likely to emerge only through continued social struggle within the contested terrain of dryland forestry, as conflicting demands and interests are constantly renegotiated.

8

Soil Erosion ▌ Breaking New Ground

MICHAEL STOCKING

Introduction

Soil erosion is one of life's certainties. It happens. Absolute quantities of soil move from one position in the landscape to another or to a river and thence to the sea; we measure how much is lost, we know what is the impact, and we have an armoury of techniques to stop it. An acceptable annual loss of five tonnes of soil per hectare may be contrasted, for example, with an unacceptable 10 tonnes. Gullies are caused by erosion; erosion is bad; therefore gullies are bad. Vegetation, particularly nice thick forest, is good; bare soil is bad; therefore, vegetation should be planted to protect the soil from erosion. These, along with others, are accepted wisdoms of theoretical, experimental and observational science. They seem to present familiar and reassuring facts. For the most part, such facts and the conclusions to be derived from them about soil erosion (and its corollary, the need for soil conservation) are well-founded within the bounds of statistical and measurement error.

Sometimes, however, there is a hidden agenda in which it is convenient to accept unquestioningly the results of natural science. But why question the output of science? First, it can simply be wrong. This can happen in a wide variety of ways, the most common of which are through the acceptance of false assumptions, particularly about scale effects, the degree of generalization, and the capability to interpolate and to extrapolate. Measurement can go awry. The typical means of measuring rates of soil erosion may be so intrusive that it has a greater influence over the result than the process itself does. A further way of getting wrong results is through over-simplification of interactive variables; for example, the interaction of rainfall and vegetation in controlling soil loss. The role of vegetation, as will be described in this chapter, is extremely complex, affecting soil condition, infiltration, microclimate, the partitioning of water flows, and the interception of raindrops. These are all difficulties which could be overcome in the measurement of soil erosion, except that to overlook them will often conveniently exaggerate erosion, and therefore prove useful to certain professionals and policy makers.

This suggests a second reason for questioning the output of science. Scientists are just one set of actors in the 'soil erosion game', a game in which it is advantageous (a) not to admit you do not know the answer; (b) to make unverifiable assumptions so that, if your answers provide bad advice, blame does not attach to the professionals; and (c) to exaggerate the seriousness of the process in order to gain kudos, prestige, power, influence and, of course, further work. In short, soil erosion 'facts' may be as hidebound with bias, error and prejudice as the outpourings of social science. The following sections will show how science could give wrong or partial or misleading results. A major pillar of received wisdom in narratives about environmental change and environmental policy-making in Africa may not be as solid as some scientists and policy makers would wish us to believe.

Images of soil erosion in Africa

Soil erosion evokes emotive images for many people and encourages others to use strange but vivid metaphors to illustrate its seriousness. A government minister in a newly independent African country, describing the severity of erosion in the communal areas, said that sufficient soil was being lost annually to fill two goods trains stretching between the earth and the moon. Are such images true? Probably not. The figure of total tonnes of soil lost from which the minister did his 'back-of-the-envelope' calculation was derived from field soil-loss rates from 30×10 m erosion plots, extrapolated to large catchments, and then summed for the whole country. As will be described later, this is likely to give figures of total sediment loss of at least two orders of magnitude exaggerated. The goods train and the distance to the moon made a good story.

Typical scenarios that are drawn of environmental degradation in Africa will usually include one or more of the following features: huge, canyon-like gullies, with bare and collapsing sides caused by waves of sediment-choked runoff; 'moonscapes' of stones, or treeless slopes littered with the debris remaining after erosion; or the remains of a once-pristine forest, the blackened stumps still giving wisps of smoke, with the soil baked hard into nodules of brick.

Along with such purely physical evidence of the abuse of the landscape are frequently evoked human images that suggest: here are the guilty; 'we' the scientists or policy makers know better; and your leaders are an abject bunch to let you do that. So, we are presented with: a peasant weeding around a stunted plant of millet growing in a pocket of soil between boulders on a 100 per cent slope [subconscious message: surely, even you could work out that this is unsustainable]; an emaciated cow, head down, on a field of sand [message: fancy keeping

that many cattle on this land – your values and customs are to blame];
and a dejected group of villagers standing around a dried up waterhole
[message; well, you had it coming to you and now you've found out
what abusing your environment can do].

The converse image whereby soil and water conservation are
promoted is also in vogue. Typical scenes might include: singing,
smiling work gangs of (usually) women, digging terraces [message: see,
it can be fun]; beautifully trimmed parallel hedgerows with closely
spaced intercrops of beans and cereals with a (usually) male farmer
holding a sample cob of maize [message: look, if he can keep a
manicured farmscape, so can you]; and a chunky Friesian cow, being
stall-fed luscious planted grasses and legumes [message: really, you're
missing out on all that money and milk].

Of course, these images are well-intentioned. However, they deliberately set the contrast between conservation and erosion; good and bad;
wisdom and stupidity; education and ignorance; innocence and guilt;
technology and local practice; professionalism and guesswork. As such,
they are neither accurate nor helpful, and they could serve to mislead.

In the following sections, six commonly held conventional wisdoms
concerning soil erosion and conservation based upon measurement and
science are examined for their potential to mislead. Particular emphasis
is laid on how science may get it wrong. It is left for others to ponder
the degree of duplicity involved, and to other chapters in this volume to
explore the political and institutional contexts in which such wisdoms
are useful, and in which they are produced and sustained.

Gullies are the worst

My first encounter with the deceptive power of gullies came as a result
of over five years monitoring erosion rates in central Mashonaland,
Zimbabwe, in the early 1970s. Led to the area first by a worried
agricultural officer, I looked at the gully erosion processes with the view
to understanding how these canyon-like forms could be controlled. It
was clear that these gullies were ignoring all attempts to halt their
apparent progress upslope. Stone gabions were being left high-and-dry
in the middle of the gully channels and the protection of the gully heads
by vegetation and stones was being destroyed during each major storm. I
measured not only the movement of gullies and their net contributions
to total sediment but also the rates of sheet wash in the intervening
areas and sediment budgets in the rivers.

The conclusion I did not want was the one I came to. The gullies
contributed only a small part of the total sediment (about 13 per cent in
the worst areas); sheet erosion from the gully catchments provided the
rest. This research in turn led to an elaboration of threshold concepts of
gully initiation and extension, where the gully in effect is triggered by

upstream conditions such as contributing area and runoff rates, and its existence is maintained as nature's efficient conduit for getting rid of surplus water and sediment. In other words, the gully was not the erosion; it was the means to dispose of the sheet erosion in the catchment. It was little wonder that conservationists' efforts to protect the gully within the gully failed: they were addressing the wrong problem in the wrong place.

Responses to orthodox wisdom about gullies can be illustrated with reference to two East African countries. In one, its President, a committed conservationist, is well known for his propensity to halt his motorcade whenever he observes a gully and to order his retinue to throw rocks and trees into the channel. The scene makes a good image for the local media. The second illustration concerns the Malanje Gully on the outer side slopes of Ngorongoro Crater. This famous scar on the landscape conveniently faces the traveller as the main road descends from the crater edge and makes its way down over the Serengeti Plains to Mwanza and the shores of Lake Victoria. The blackness of the *mbuga* soil exposed within the gully contrasts with the green of the short grass plains around. Dignitaries, foreign and local, pass that way, and the Malanje Gully is an acute source of embarrassment to the Ngorongoro Conservation Area Authority. The Maasai get the blame, as the gully is used as a path to herd cattle to water. However, which came first: the Maasai track which formed a gully; or the gully which provided a useful track? The retired Conservator of Ngorongoro, Henry Fosbrooke, has some good photographs of the gully from the early 1950s, and detailed comparative surveys have been carried out of the gully then (from the photos) and today (in the field). These indicate almost no change in its position and extent: indeed, if anything, parts of it are healing and slowly revegetating. The case against the Maasai is thus very shaky.

What, then, can we conclude about gullies? They have a bad press. As evident features of the landscape, their 'guilt' is obvious. Yet, most evidence indicates that they contribute only modest amounts of sediment; that they are a symptom of a degraded catchment, not the cause; and that they are a response to degradation elsewhere, rather than in and immediately around the gully. They can, therefore, be used as a political weapon as demonstrated in Ngorongoro against the Maasai; yet as Homewood and Rodgers (1984) have argued, there is very little evidence to associate either the invisible sheet or the obvious gully erosion with the activities of pastoralists. Unlike gully erosion, sheet erosion is insidious, rarely visible without close examination. Total quantities of soil loss through the removal of discrete layers are far greater even where sheet and gully erosion exist together. Sheet erosion's impact on site productivity far surpasses that of gullies; experiments at the International Institute for Tropical Agriculture (IITA) in Nigeria, for example, showed that 75 per cent of maize yield is lost with only 1 cm of soil loss (Lal, 1976). A gully may render its immediate area totally unproductive but this would typically be less than 1 per cent of the total landscape.

Stones everywhere - must be serious erosion

Some landscapes are littered with stones. Östberg's (1991) account of the Burungee revealed fascinating explanations as to how and why 'the stones are coming up'; they regard stones as evidence of soil formation, not erosion. While undertaking an erosion survey in Shinyanga Region, Tanzania, I came across extensive areas of Maswa District above Lake Eyasi where only a bare rubble of calcrete covered a surface where once, having witnessed it from aerial photographs, I knew *Acacia - Commiphora* woodland to have existed. The orthodox response is to interpret present soils as remains of severe erosion caused by the cutting of the woodland and excessive grazing: only stones remain because they are too heavy for runoff to transport. This in turn implies that erosion is continuing at a fast rate.

Leaving aside the possibility that the woodland hid the existence of surface stones on the photographs and that little has really changed other than the removal of some thorny and obstructive trees, how could the stones have been derived and what is their current influence on erosion rates? Some soils naturally contain a high proportion of stones, gravel and coarse sand. The lake-bed sediments in Shinyanga consist of silty-clays, dark in colour and rich in nutrients, interspersed by calcrete nodules, almost white in colour. The nodules could take up some 50 per cent of soil volume, yet the soil would be fertile and quite attractive for cultivation. From what would have been a grey-black surface, only a few millimetres of erosion exhume the white stones. In other words, under specific conditions it takes only a small amount of erosion to create the effect of a radical change in soil quality.

A further possibility for the exhumation of surface stones is that the Burungee are actually right: stones do come up. Moeyersons (1978), working from the Royal Museum of Central Africa at Tervuren, Belgium, on sites on Kalahari Sands, has shown that wetting and drying cycles typical of savanna areas can cause both artifacts and stones to migrate upwards or downwards according to relative densities of material and the swelling and shrinkage characteristics of the soil. Apart from being a salutary warning to archaeologists to beware of dating fossils from the horizons in which they are found, the process is probably responsible for many stone-strewn landscapes in which net erosion is negligible.

The current influence of surface stones is usually to reduce erosion rather than to indicate its severity. In effect they act as a surface 'mulch', absorbing the energy of rainfall as effectively as any vegetation. The danger with stones is that they divert water to micro-channels where the water concentration might exceed the infiltration rate. On slopes the runoff yield is greatly enhanced, as shown by the work of Yair and Lavee (1976) in the Negev. However, laboratory experiments reported by De Ploey (1981) show conclusively that under certain conditions stones have an obstacle effect and that they can even migrate upslope. Stones, therefore, have conflicting effects, but there is abundant evidence throughout Africa that farmers use stone-strewn landscapes for productive cultivation; that surface stones reduce contemporary erosion; and that only modest amounts of historical erosion are needed to create apparent 'wastelands' or 'badlands'.

Vegetation protects

Manuals on soil conservation stress the importance of vegetation cover in preventing erosion splash (the initial detachment process), encouraging the infiltration of water (thereby reducing sediment transport), and providing mini-barriers against overland flow that sift the sediment and encourage redeposition. Further claimed effects of the efficacy of vegetation include its role in maintaining the quality of soil through the cycling of organic matter, the encouragement of a microclimate suitable for beneficial micro-organisms, and the physical binding effect on the soil of plant roots. Research generally supports the belief that vegetative cover is the single most important factor in soil erosion control in the tropics (Stocking, 1994). New approaches to soil conservation, the so-called 'Land Husbandry' strategies (Shaxson *et al.*, 1989; Lundgren *et al.*, 1993), are predicated largely on crop management and the cycle of organic matter as the best way of promoting soil quality and hence of achieving soil conservation.

Not always, however, does vegetation have a positive and protective influence on the site in which it grows. De Ploey (1981) included vegetation as one of his 'ambivalent factors' after observing experimentally in the laboratory at Leuven (Belgium) an increase in sediment from a trough under simulated rainfall planted with bunch grasses compared with another trough that was bare. This led him and others to observe a number of situations in which the orthodox response – to plant more vegetation to reduce erosion – could be ineffective or even counterproductive. Such situations in Africa include the following:

Bucket phenomenon: certain tree species, such as teak (*Tectona grandis*), have large, bucket-like leaves. While these are highly effective in protecting the soil surface for, say, the first ten minutes of an intense storm, the erosive power of rainfall is considerably increased thereafter, once the weight of water bends the leaf-stems. If detachment power of drops is directly proportional to kinetic energy[1] (Stocking and Elwell,

---

[1] Kinetic energy = ½*mass*velocity^2

1973), then the ability of these super drops to splash soil particles is immense.

Cover height: more generally with trees and other forms of plant cover that are well above the soil surface (such as mature maize plants), reformed droplets from leaves can regain a high percentage of their terminal velocity as well as having significantly enhanced size. This phenomenon was noted in some of the earliest work on vegetation and erosion by Screenivas *et al.* (1947) for planted crops and by Geiger (1966) under a forest canopy. A single droplet under vegetation cover could have far greater punch than a single droplet of natural rainfall. A 5 mm drop, falling from a leaf 4 m above the ground, has eight times the kinetic energy of the typical median drop size (2.5 mm diameter) of tropical rainfall at terminal velocity. Under a tree canopy, the pitting of the ground surface by splash is often clearly visible, as for example under *Acacia* spp. on the Maasai plains that are grazed to giraffe-mouth height (cf. Vis, 1986, for Colombian forests).

Plant structure: as it intercepts rainfall, vegetation partitions the water into throughfall, leaf drip, leaf evaporation and stemflow. In some plants, the latter can be dominant, especially where the structure of the plant channels water to the stem. In a gully reclamation scheme in the Ngezi area of Zimbabwe, sisal (*Agave sisalana*) was planted in rows in an attempt to stabilize the gully heads. Instead, the sisal leaves concentrated water down the stem of the plants, causing serious rilling and eventually small gullies before eventually the sisal plants themselves were washed away in a storm and the channels obliterated. Other species prone to this effect include palms, bananas, maize, and a number of the Miombo trees in the seasonal tropics of South-Central Africa.

Faunal activity: in landscapes as diverse as the open deciduous forests of the Miombo; the dense, dry forests of Zaire; and true semi-evergreen tropical rainforest, the activity of termites and other soil-moving fauna is truly remarkable. In southern Zaire I have counted some two mounds per hectare of *Macrotermes* and at least 50 mounds of smaller species of termites, which are likely to have brought to the surface around 5 tonnes/ha/year of soil. This bare soil, exposed on the steep sides of mounds, is prone to splash and transport, exacerbated by the litter fires which sweep through the dry forest virtually every year. Planting vegetation in order to reduce erosion would have little effect under such conditions.

Litter transport: in the seasonally wet tropics, large quantities of leaves and other vegetation debris are carried away by overland flow. Regularly, one can observe the organic debris of grasses, leaves and twigs at the footslopes of gullies and deposited in valleys. While probably not large in terms of total mass removed, litter transport constitutes the vast bulk of total nitrogen losses from a site and thus accounts for a substantial proportion of total soil degradation. Nutrient loss analyses in Zimbabwe from land under crops indicated an average biomass loss by erosion of 500–750 kg/ha/year dry weight, the exact level depending on soil type and erosion rate (Stocking, 1986).

Turbulence around stems: the stems of many bunch grasses, especially, alter the hydraulics of runoff, and have been seen to promote turbulent flow patterns and the activation of additional sediment transport. The laboratory studies of de Ploey *et al.* (1976) conclusively showed that on certain slopes, the concentration of sediment load on a grass-covered soil was considerably greater than on an otherwise identical bare surface. It is only when basal cover exceeds a certain threshold (probably about 15 per cent) and canopy cover exceeds about 45 per cent that vegetation regains its protective role. The majority of the savanna grasslands of Central Africa have vegetation covers below these critical levels and a poor cover of ground mulch, especially at the beginning of the rains when the most intense and erosive storms may be expected.

These rather ambivalent effects of vegetation on soil indicate a careful need to assess the specific conditions of site, slope, rainfall and vegetation type before determining whether or not plant cover is likely to conserve soil. Vegetation communities may either decrease or increase erosion rates. The balance between the two represents the net effect, but this changes substantially as vegetation is either removed or replanted. Two other aspects of vegetation need to be considered when assessing orthodox responses to erosion: allelopathy and the 'Eucalyptus debate'; and the use of tree pedestal heights to assess erosion rate.

In the late 1960s and early 1970s, in Southern Rhodesia, I noticed how one set of gullies appeared to have changed headward direction towards a plantation of *Eucalyptus grandis*. These were magnificent trees, 30 or more metres tall, planted two decades earlier as a demonstration by local forestry staff of how to conserve erodible soils. Plotting the headward advance of the gullies from time-series aerial photography, it seemed deliberate that the gully should have turned to threaten the trees that should by rights have been protecting the landscape. Today, the gully runs through the plantation, and many of the trees have collapsed into the channel with their roots exposed.

This was my first exposure to the controversy surrounding Eucalyptus. Opponents of the tree claim that allelopathic interactions with other species mean that ground cover is prevented. Indeed, many examples abound where Eucalyptus and other species such as pine seem to stifle the all-important grasses and herbs that do prevent erosion. Allelopathy need not, however, be invoked. In comparing two plantations in somewhat similar climates in Lesotho and Malawi, it is clear that in some instances the ground is absolutely bare under Eucalyptus while in other places a dense undergrowth can co-exist with the tree. The only obvious difference is the quality of the soil and its ability to supply the necessarily large water requirement. An already degraded soil, on which Eucalyptus is popular because of its ability to withstand harsh conditions, will only be able to supply the tree and no other vegetation, whereas good soils with a high plant-available water capacity can support both. This ecologically demanding aspect of Eucalyptus has been reviewed by FAO (Poore and Fries, 1986); it means that site condition, especially the degree of prior erosion, must be considered carefully before the tree is used for conservation purposes.

The second additional vegetation consideration that requires extreme caution is the use of soil mounds beneath trees to indicate the quantity of erosion (or ground lowering generally) since the trees commenced growing. Pioneered in semi-arid Kenya by Dunne *et al.* (1978), the technique was adopted in the early 1980s for an erosion survey of Shinyanga Region, Tanzania (Stocking, 1984). After using the method in the Hardveld of Botswana, however, Biot (1990) believes that the results obtained are untrustworthy, since the mound under a tree could be explained by factors other than erosion. Lowering of soil-bulk density in the direct vicinity of the tree stem and roots because of higher organic matter content, decaying root material, and termite activity, could have the effect of raising the local surface. Biot's research also suggests that trees in semi-arid areas encourage deposition of sediment at their base, thus enhancing the differential heights of the soil surface between base and intervening areas. If he is correct, then calculated measures of erosion rate over 10 to 50 years (the typical ages of trees as measured by tree rings), including some of my own calculations, could be grossly exaggerated. Support for his conclusion comes from erosion estimates using the soil loss model, SLEMSA, of about 5 to 8 tonnes/ha/year (Abel and Stocking, 1987), whereas tree-mound heights were giving estimates averaging 90 tonnes.

Erosion rates are meaningful

Erosion rates usually refer to the assumed total loss of soil from a site, represented as tonnes of soil per hectare per year. Geomorphologists often refer to essentially the same measure but in units of millimetres of ground lowering per year. One millimetre of soil lost is the same as 13 tonnes/ha/year (assuming a soil bulk density of 1.3 $g.cm^{-3}$). The single most common means of deriving such rates is the field soil-loss plot, a bounded area usually about 20 m long upslope and 2 to 3 m wide across the slope. At the bottom of the plot a trough collects the soil and water that is washed into it. It seems convenient to set up a representative sample of these plots for typical land uses, and use the results to make estimates of catchment erosion (say, for danger of reservoir sedimentation), and of regional or national erosion (for computation of broad environmental impacts). Indeed, the vast majority of countries in Africa have at some time had massive erosion research programmes based upon field plots (Stocking, 1992). Notable among these were the programmes in Central Africa for the Federal Ministry of Agriculture (Hudson, 1957) and West Africa for ORSTOM (Roose, 1975).

What fundamentally is wrong with soil-loss and runoff plots at the field scale? In their defence, they are excellent at demonstrating the differential effects of land uses and hence the danger of planting some crops in particular ways. In Southern Rhodesia in the early 1970s, for instance, when used to compare the effect on erosion of planting pattern and planting density of cotton at Henderson Research Station (Stocking, 1972), they revealed a clear difference between the high-density/high-fertility plantings and the low-density planting typical of communal areas on sandveld soils. In order to make conclusions about relative rates of erosion and relative impacts of different land uses, then, field erosion plots are extremely useful. Unfortunately, most researchers have gone further. The driving force was the search for factor values in the Universal Soil Loss Equation (USLE). In the United States by the late 1960s where the USLE was *de rigeur*, some 3,000 plot-years of data had already been assembled by the Agricultural Research Service. In turn, African researchers also set about assembling a USLE data base of absolute values of soil loss. The problem with the USLE, as with any other empirical model, is that it is data-demanding and totally dependent on the accuracy of the input; in this case around six parameter values. In Africa at least, the dubious validity of data means this was an unrealizable goal.

Proposals for major measurement programmes of soil loss using field plots still appeal. What is wrong with them? First, absolute rates are not particularly helpful. For reasons elaborated below, they say little about what sediment is transferred to another site, say a reservoir. More importantly, the impact of erosion on productivity is both soil- and plant-specific. One centimetre of erosion may cause yields to crash on a very susceptible soil (a Luvisol, for example), yet have little effect on a well-drained, high fertility clay (a Nitosol), and may even cause yields to increase on another soil (e.g. a duplex soil with greater exposure of clays with better water capacity). The erosion-productivity linkage is exceptionally complex. But what can be said definitively is that absolute rates of soil loss are useless for differentiating the impact of erosion between soils and between crop types.

Second, erosion plots also provide a classic case of experimental interference, the phenomenon whereby measurement itself intrudes on the process being measured. Erosion plots set up impermeable boundaries that (1) prevent natural overland flow from upslope, as well as imports and redepositions of sediment within the plot; (2) alter the balance between detachment and transport processes, primarily through unrealistically limited overland flows; (3) interfere with rilling and sheet erosion down the long boundaries; and (4) create artificial land uses because of the difficulty of squeezing real practices into a small plot. Consequently, the processes being measured in and from the plot are extremely unlikely to resemble those on a real field.

Third, experimental errors are also likely. Sediment and runoff pass from the trough to a 'sludge' tank. The usual method of calculation of soil loss is to stir the tank and take a sample of suspended sediment and water; dry it in the laboratory; and then calculate total loss from a knowledge of the depth of water and sediment in the tank. After some discrepancies in replicate samples, Henry Elwell at the Institute of Agricultural Engineering in Harare compared results between the normal sampling technique and the extremely laborious drying and weighing of the total sludge. He found a gross underestimation of actual soil losses by the sampling technique, presumably because of inadequate mixing. He further showed that, however hard one stirred, it was virtually impossible to get a representative sample. Clays would be over- and sands under-represented in the sludge samples, while vigorous stirring still gave some 10 per cent less soil loss than total sampling. He devised a total weighing technique to overcome the problem on all the Zimbabwean sites, but I am not aware of any other research groups having adjusted their techniques because of these findings.

Fourth, erosion plots also suffer from scale effects. An erosion plot only catches sediment that passes into the trough and thence to the sludge tank. It does not monitor soil particles which start to move within the plot but redeposit themselves before reaching the trough. This produces a scale effect analogous to experimental interference. Small plots have a greater chance of catching all soil particles which start to move; while large plots have many particles which start to move then stop, and real slopes have a balance of erosion and deposition throughout the length. Consequently, on a per hectare basis, small plots suffer higher soil loss than large plots, which in turn may (depending on the importance of overland flow from the whole slope) be very much higher than a real slope. Hydrologists call this scale effect the Sediment Delivery Ratio (SDR), meaning the sediment that exits the catchment in proportion to the field soil-loss rate. SDRs for small catchments are typically 0.1; for major catchments, 0.05; and for large river systems, 0.01. Therefore, to extrapolate regional- and continental-scale erosion from field-plot results invites an exaggeration of actual sediment loss by a factor of 100 or more.

What is the alternative to erosion/runoff plots? As noted earlier, research resources were diverted from the 1930s or so in response to the mid-West US Dust Bowl to measure absolute rates in order to provide factor values for the Universal Soil Loss Equation. The direction of soil erosion research in Africa was greatly affected (Stocking, 1992). Implicit in that diversion is that somehow soil-loss rates are related to the impact of erosion on soil productivity, crop yields and livelihoods. Erosion rates are, however, only poorly related to these. Hence, the obvious alternative would be to return to the original impetus for erosion research: the effect on productivity. In practical terms, this would involve land-use and farming systems analysis in relation to biophysical sustainability indicators such as nutrient losses and organic matter depletion, and economic indicators such as cost-benefit measures (e.g. Net Present Values), which would determine the farmer rationality of investing in soil conservation. Work recently completed in semi-arid Kenya adopting this approach showed conclusively that the out-of-fashion conservation technique of trashlines was by far the most profitable for a small farmer (Kiome and Stocking, 1993). Surely, it is better to by-pass a physical measure of dubious validity – absolute soil-loss rates – and go straight to measures of immediate applicability for land-use decisions, such as crop yields or returns on investment.

Everyone suffers when erosion occurs

It is conventional to assume that all erosion is bad, and that everyone suffers from a depletion of natural resources. Erosion affects site productivity, then on its way downstream it chokes water storage facilities and navigation channels; it damages hydro-electric power installations; it sediments ports; and finally it pollutes off-shore waters, damaging marine fish resources. A picture could be drawn in which everyone loses because of the damage created by erosion. This assumption needs close examination, however. First, because soil and sediment do have an intrinsic value in nutrients and productive potential, which could conceivably be transported along with the soil. Second, because if erosion is such a loser, surely measures would have been taken by now to prevent it. Third, because it would appear that some people, usually small-scale farmers and peasants, are acting irrationally in causing erosion which could only be to their detriment. In brief, people are not acting in a way that suggests they feel erosion is harming them.

A case from the hill lands of Sri Lanka shows the transfer of productive benefit from one site to another by erosion. In the late 1970s, the UK Overseas Development Administration funded the Victoria Dam, part of a complex of five major reservoirs to supply electricity and irrigation water to the Mahaweli scheme. A worry at the time was the life-span of the reservoirs, given the cultivation of steep hillslopes in the catchments already, the displacement of farmers from flooded areas, and their migration to the hills to cause yet more erosion. Despite the worries and half-hearted attempts to control cultivation in the catchments, the Mahaweli scheme went ahead. 1991 and 1992 proved to be relative drought years for the hill lands and the level of water in the dams decreased sufficiently to survey some of the water storage beds. It transpired that sedimentation was negligible; instead of the predicted deep thicknesses, there were only minimal layers of fine clay. Yet, looking up into the hills, the signs of erosion were everywhere. Small farmers are today growing tobacco, a notoriously dangerous crop because of its poor cover, on steep slopes; vegetables for the local market are also evident, requiring much cultivation and weeding and hence soil displacement. But that soil was not arriving in the dams downstream. Where was it going?

Many of those same farmers on the slopes also had rice paddies at the base of the steep slopes. Traditionally they dig earth bunds to pond back the water, and most would get at least one rice crop and probably a dry season crop of pulses or chillies. It was fascinating to discover that they never used inputs on their paddies; that they were investing heavily their own labour in building more paddies, some of them starting to impinge on the steeper slopes; and that most of their effort went into these fields, rather than the quick cash crops on the slopes. Literally, in the paddies they were farming the soil of the slopes along with its nutrients and organic matter. It was a good illustration of erosion as a game of winners and losers, except that in this case, for the farmers with plots on both the slopes and bottom land, the balance of benefit from erosion was positive; it was advantageous to use erosion as free transport of inputs to their most valuable crops, the rice and pulses.

One good measure of the intrinsic value of sediment (and also the detrimental loss to the site of erosion) is the enrichment ratio. Erosion is selective, removing the finer particles including clays and organic matter. Hence, in nearly all cases, sediments contain on a volume-for-volume basis far greater amounts of nitrogen, phosphorus, other nutrients and organic matter. The enrichment ratio (ER) is simply the proportion by which the sediments are 'richer' compared to the soil from which they were derived. Typical ERs from long-term experiments in Zimbabwe were 2.5 for nitrogen and phosphorus (Stocking, 1986). In Ethiopia, monitoring of ER on a clay-rich Nitosol gave considerably lower values (Tegene, 1992), highlighting the more general phenomenon that the impact of erosion on poor soils is proportionately far greater than the impact on good soils.

In more general terms, erosion represents a redistribution of soil nutrients and plant production potential. In aggregate, it is accepted that a net loss to society probably occurs more often than not, but this loss is made up of some who gain and some who lose. The gainers could be other farmers using the sediments directly, as in the Sri Lankan case; they could be land users that society now subsidizes to keep on the land or to construct conservation measures; or they could be politicians benefiting from a peasantry kept poor.

Soil conservation is the answer

The term 'soil and water conservation' describes a large number of techniques and measures designed to prevent erosion from starting, or to control the process once commenced. Prevention is usually achieved through the protective role of vegetation (notwithstanding its ambivalent effects noted above) and control is exercised by structures and mechanical means that encourage sedimentation and channel water away safely.

Just as it is assumed that erosion is bad, conservation is conventionally considered to be good. Just as the previous fallacy has it that all suffer from erosion, this myth assumes that all will benefit from conservation. However, the rationality of conservation is far from simple, as can be demonstrated from an analysis of the degradation of a humid tropical Oxisol in Sierra Leone (Sessay and Stocking, 1993). Erosion to different degrees had notable effects on the soil's physical properties (e.g. increased bulk density, reduced infiltration, decline in pH status and plant available water) and chemical properties (reduced organic matter and decrease in total supply and rate of availability of nutrients). These influences worked in tandem to make crop production less worthwhile. However, the analysis also predicted the benefits of reversing the process through conservation. At least in the short term, it showed that investment in terraces and other structures, designed to reduce sediment yield, would have no economic return to the farmer. The same observation has also been noted under quite different environmental conditions in semi-arid Kenya (Kiome and Stocking, 1995).

Conclusion

This chapter has deliberately sought to expose the contradictions between the scientific evidence of erosion, the selective use and misuse of results, orthodox 'knee-jerk' responses to visual evidence of erosion, and policy responses. The 'myths' referred to concern particular physical phenomena rather than whole policy discourses, and hence they are somewhat different to the received wisdoms discussed in other chapters of this volume. However, these physical phenomena are the foundations and building blocks of rationalizations for many of those broader policy orthodoxies. Once such observations are cemented together by particular theories of environmental change and supported by political and institutional pillars, they provide compelling 'evidence' for the wisdom of many actions in the arena of agricultural and rural development in Africa – such as the current trend for writing National Conservation Strategies as in Namibia, Zambia, Botswana, Ethiopia and other countries.

Of course, in many places erosion is bad; crop yields are crashing; land is being abandoned; people are migrating; and a cycle of degradation is commencing on marginal lands as population pressure is transferred. However, to claim that these represent universal links of cause-and-effect is far too simplistic. The evidence from Machakos district in Kenya (Tiffen *et al.*, 1994; cf. Tiffen this volume) shows indisputably that erosion may be a passing phase in a farming landscape, during a transition to different land uses, higher populations and new technologies. In these terms, erosion may be good, if it forces populations to adjust. Yet the orthodox response to the photographs of 1937 Machakos would be 'do something', 'bring in the bulldozers' or 'tap World Bank environment funds for a project'.

Insofar as they encompass received wisdom and a selective use of science, the six myths of this chapter need addressing. The myths can be seen as the result of simplistic assumptions, visual responses and occasionally measurement error and misjudgement. But they need to be set in context: first, of the experiment that may have determined a result, including the type of measurement, scale, and possible errors. Second, of raising questions about the circumstances under which different sorts of experiment are deemed appropriate, the selection and rejection of scientific results by scientists themselves, other professionals, casual observers and policy makers. And third, most difficult of all, of the interconnections between supposed cause-effect linkages and other processes, and the prejudices of observers who might prefer to see something bad than something innocuous. After all, environmental policies are much more convincing if one can argue that an environmental catastrophe is imminent.

9

Irrigation, Erosion & Famine

Visions of Environmental Change in Marakwet, Kenya

W. M. ADAMS

Ideas, indigenous knowledge and irrigation[1]

It is now conventional to recognize the enterprise and ingenuity of small farmers and other resource users in sub-Saharan Africa, and the 'indigenous technical knowledge' on which they are based. Writers commonly now stress both the skill with which farmers manage, or adapt to, their environment and the massive importance of that skill and adaptiveness to any successful attempt to produce effective developmental change. This enthusiasm for indigenous knowledge (IK) became fashionable through the 1980s, building on the work of people like Robert Chambers and Paul Richards (e.g. Brokensha *et al.*, 1980; Chambers, 1983; Richards, 1985). It has now become a received wisdom in development thinking, at least among Western aid donors and researchers in the Third and First Worlds.

At its best, this new thinking offers a radical and emancipatory challenge to conventional high-tech top-down development planning. At its sloppiest, it represents a flabby, untheorized and romantic view that everything farmers and rural people do is wise, fair and sustainable, and that it has remained unchanged since the dawn of time, until developers come along with Western culture and technology to upset the natural balance between people and nature.

In the specific context of African irrigation, a new awareness has grown up within the last decade of the diversity and extent of

[1] The research on which this paper draws was carried out with Elizabeth E. Watson (University of Cambridge) and in collaboration with Dr S.K. Mutiso of the University of Nairobi (Department of Geography, Kikuyu Campus). Research permission was obtained from the Office of the President (Permit No. OP 13/0001/9c 227/16). It was funded by the Economic and Social Committee on Research (ESCOR) of the Overseas Development Administration (ESCOR Project R4760, 'Socio-economic Aspects of Technical Intervention in Farmer-Managed Hill Irrigation, Kenya'), and the Economic and Social Research Council (ESRC, Award No. R 000–23–2932, 'Socio-Economic Change and Sustainability in Indigenous Irrigation, Kenya'). The author would like to thank John Sutton for many valuable insights on the history of irrigation in East Africa.

indigenous practices of gravity and groundwater irrigation and wetland agriculture in sub-Saharan Africa. Research has started to reveal something of the history of these irrigation systems, and the social and environmental contexts in which they have functioned over time (e.g. Sutton, 1984, 1989; Anderson, 1989). The more generalized ideas about IK have been reflected within debates about irrigation and development. It has been argued that these systems have been inadequately studied and understood, that in many cases they represent sound and sustainable ways to manage and develop land and water resources in sub-Saharan Africa, and that formal modern development projects have not only ignored them but in many cases directly disrupted or dislocated their functioning (e.g. Adams, 1992).

Both because of the failure of the formal large-scale irrigation schemes built over the last two decades and the groundswell of enthusiasm for small-scale, farmer-managed and indigenous irrigation systems, a body of work has emerged that focuses on the potential for development projects building directly onto existing indigenous farmer-managed irrigation schemes (Adams, in press). These projects variously seek to engage directly with existing irrigation systems, whether identifying and implementing technical improvements (often called 'rehabilitation' projects, as in Tanzania), or intervening in social organization (for example establishing farmer-management associations; Cernea and Meinzen-Dick, 1994), or through intervention in the organization and control of labour and land tenure (as, for example, in the Gambia; Carney, 1992).

However, this recent upsurge of interest by outside development agents in existing irrigation systems is not new. Former generations of powerful and sometimes technically trained outsiders have come upon these irrigation systems and pronounced upon them, in the light of their outside-generated visions for change or development. Sometimes these outsiders have also sought to intervene in these systems, and proposed that they should be changed, just as their contemporary successors are doing today. Such intervention represents the imposition on the landscape of contemporary thinking about development.

Ideas about development change, moving in and out of fashion. As they do so, prescriptions for development change also, reflecting shifts in the received wisdom of national government bureaucracies and international 'expert' debate. At any one time, a particular set of ideas will tend to dominate development thinking at the centre, and new projects on the ground will reflect these ideas. However, this received wisdom will not be uniformly accepted everywhere, and may even be contested as contrary narratives flow back from periphery to centre (cf. Roe 1991). There will certainly be lags in the acceptance of new thinking as received wisdom, and these will be reflected also in space as ideas move from planning room to field project. This paper describes the way in which change in received wisdom about indigenous irrigation on the Marakwet Escarpment in the Kerio Valley in Kenya influenced development interventions on the ground. It explores in particular the views of colonial officers and development agents between the 1930s and the 1960s, as recorded in the files of the Kenyan National Archives.

Irrigation in the Kerio Valley

Irrigation takes place by gravity along about 50 km of the western escarpment of the Rift Valley in Kenya above the Kerio River (Figure 9.1). The system of hill furrow irrigation is similar to that in a number of sites in East Africa, for example in the Taita Hills, Sonjo in Tanzania, and the slopes of Mount Kilimanjaro (Fleuret, 1985; Adams *et al.*, 1994). It is also technically comparable with the extinct late-Iron Age irrigation system at Engaruka in northern Tanzania (and with modern gravity systems there and in many other places in East Africa).

The irrigation system is conceptually and technologically simple, although it exists in dramatic country and covers such an altitudinal range and such great distances that it represents a remarkable human achievement (Soper, 1983). A series of streams rise high on the Cherangani Plateau (originally in the evergreen forests, although these are now increasingly cleared for agriculture), and descend the steep Marakwet escarpment to the valley floor some 1450 m below, where they flow eventually into the River Kerio. There is some 1500 mm of rainfall on the plateau top (falling in two rather variable rainy seasons), and only 600 mm on the valley floor. Water is diverted from these streams using simple dams of brushwood and stones into channels that are built of earth and stones. These are led along the slope and steadily down it, and the water is used both to supply houses and to irrigate fields on the escarpment side, and to irrigate fields on the valley floor. The natural streams are steep and rocky, and have highly variable flow regimes. The irrigation furrows are also steep, and use both simply constructed but ingenious engineering structures (e.g. aquaducts made of wood, mud and leaves) and natural features of the hill (e.g. bare rocks as waterfalls) to carry the water to its destination. Increasingly, these engineering structures have been constructed with modern materials (cement and corrugated iron and pipes of various kinds). There is quite a complex system for the repair and reconstruction of furrows, and for the allocation of irrigation water (Ssennyonga, 1983; Adams *et al.*, in prep. b). Residence and land tenure is basically organized on a clan basis, and this in turn dominates the water allocation system.

By the 1930s the furrow system on the Marakwet escarpment was quite well known to colonial administrators. A report in 1942 wrote of 'the famous system of furrows in Marakwet, many years old and constructed by the natives themselves' that took water from the rivers

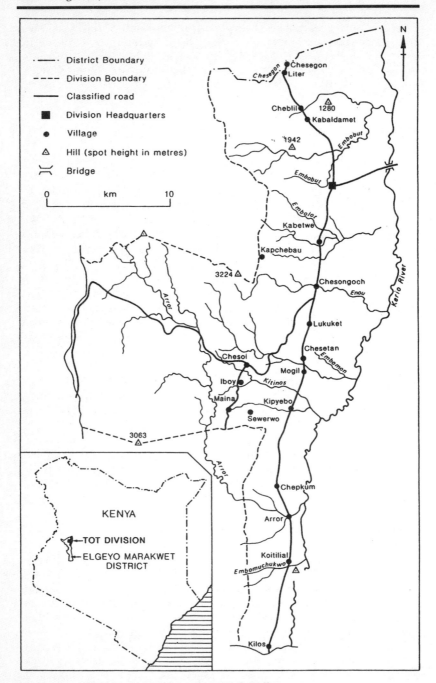

Figure 9.1   The Marakwet Escarpment, Kerio Valley

Arror, Embamon, Embobut and Embolot 'many miles along the sheer face of the escarpment to irrigate shambas on the level land below'.[2] The first outsider to note the irrigation system was Joseph Thomson, in 1885. From then on, Provincial and District records are full of comments on the irrigation and the Marakwet themselves. There was, however, a curious confusion in the thinking of the colonial officials. In part this reflects changes in personnel and in part the arrival of new and dominant modes of thought – new 'received wisdom'. Three such themes are discussed here: irrigation and fear of famine, irrigation and soil erosion, and irrigation and agricultural development.

Irrigation and famine

Colonial observers were aware of the importance of the furrow system to food production, and in particular its role in preventing famine. In 1939 the DC wrote that 'The Marakwet have practically no cattle but possess furrows which can irrigate their shambas. The Marakwet reaped a fair crop of *mtama* [sorghum] and groundnuts though the *wimbi* [finger millet] crop throughout the District was a failure.' In 1938 it was noted that 'the *mtama* crops in the irrigated shambas in Marakwet are coming along well', and following a visit in 1940 it was asserted simply that 'The Marakwet have been saved by their furrows'.[3]

This perception, that the Marakwet were, by virtue of their irrigation, somewhat insulated from the worst impacts of famine, was offset and increasingly replaced by a far more disparaging view of the irrigation system and the Marakwet themselves. As elsewhere, an attempt was made to introduce new 'famine crops' in Elgeyo in the 1930s, notably cassava. The attempt was abandoned because of attacks by porcupines, but the principle was established, and the Provincial Commissioner urged the District Administration to seek a food supply against famine: 'Natives now grub up roots in the forest. I should like to see these roots at their doors. Roots supplement crops subject to drought or flood.' If necessary, a 'porcupine campaign' should be mounted.[4] In 1932 the Local Native Council purchased seed potatoes for planting and paid for the transplanting of cassava cuttings and banana suckers supplied by the Agriculture Department.[5] In nearby Kamasia Reserve, seeds and cuttings for an amazing array of crops were distributed in 1932,

---

[2] DC Tambach to PC Rift Valley Province 7/1/42 with 'Report of Dams: Construction and Maintenance of, 1/1/42'; (DC/TMB/2/2/16, 20).

[3] DC E-M to PC RVP 17/6/38, 'shortage of food in Elgeyo Reserve'; DC E-M to PC RVP 26/12/39; DC Tambach to PC RVP 2/2/40; (DC/TAMB/1/2/16 Famine Reports and Food Supply, no folio).

[4] DC E-M to PC Nzoia Province Eldoret Food Crops 3/9/30; PC to DC Tambach 9/9/30; (DC/TAM/2/1/1 General Development, 6).

[5] DC E-M to PC Nzoia Province Eldoret 8/12/31 'Development in Native Reserves'; (DC/TAM/2/1/1 General Development, 13).

including maize, millet, sorghum, cowpeas, pigeonpeas, Canadian Wonder beans, buckwheat, cassava, yams, cabbage, marrow, sweet potato, pumpkins, groundnuts, rice, and cotton.[6]

In addition to new crops, it was argued that new cultivation implements were needed, as indeed was a much more enterprising and technically efficient approach to farming. Far from remarking on the sophistication of Marakwet irrigation, the DC wrote 'it is a constant source of wonder that any seed planted after the customary hen-like scratchings have been carried out ever reach maturity'.[7] Similar rather scathing views were held about the irrigation system: 'So far as I can judge, the Marakwet have not advanced one whit (apart from the cultivation of maize and potatoes) since the time when their more energetic and enterprising ancestors dug the furrows that are in use today.'[8] Despite the evidence that irrigation was important, the DC suggested that 'crops are grown simply in order to stave off starvation, a tedious duty as it were, to be relegated as far as possible to the weaker sex, and not regarded as a means of earning a livelihood.'[9]

The slight paradox of the obvious usefulness of the irrigation furrows in the context of famine, and the apparently feckless way in which irrigation and cropping were carried out, led very directly to the notion that the government should be involved in extending and improving the irrigation system. The DC toured Elgeyo Marakwet District in 1931 and 'took the valley people to task for their lack of enterprise', and was rewarded by a promise that in three Locations 'the Elders would take steps to investigate immediately the possibility of extending the system of irrigation.'[10] He planned to revisit in February 1932 and supply *posho* from famine relief funds to help with construction at a time when food was expected to be short, but he added that 'it is one thing to make these folk dig a furrow and another to make them put it to good use.'[11]

Perhaps unsurprisingly, it also proved much harder than expected to build new furrows that worked. District policy was to help in the construction of new furrows with tools and sometimes food (*posho*),[12] although it is clear that there was also considerable technical involvement on some occasions. Thus the Reconditioning Officer aided with

[6] Report by P. Booth on Kamasia and Njemps reserves 18/1/33; (DC/TAM/2/1/1 General Development, 22).

[7] DC E-M to PC Nzoia Province Eldoret 8/12/31 'Development in Native Reserves'; (DC/TAM/2/1/1 General Development, 13).

[8] PC Nzoia Province to Agricultural Officer Kitale 5/1/33, 'Native Agriculture – Elgeyo District'; (DC/TAM/2/1/1 General Development, 15).

[9] DC/TAM/2/1/1 General Development; DC E-M to PC Nzoia Province Eldoret 8/12/31 'Development in Native Reserves'.

[10] DC E-M to PC Nzoia Province Eldoret 8/12/31 'Development in Native Reserves'; (DC/TAM/2/1/1 General Development, 13).

[11] DC E-M to PC Nzoia Province Eldoret 8/12/31 'Development in Native Reserves'; (DC/TAM/2/1/1 General Development, 13).

[12] DC Tambach to PC Rift Valley Province 7/1/42 with 'Report of Dams: Construction and Maintenance of, 1/1/42'; (DC/TMB/2/2/16, 20).

work on a major furrow for Sambirir people off the Arror River. Work in 1937 was slow, the level of engineering design was crude, while the level of labour investment was large.[13] A hundred yards of a passage six feet wide had been built, but it was estimated that completion would take more than seven months. This furrow broke in 1944, and its repair was judged 'too large a job for the District Commissioner to tackle while single-handed'.[14]

Irrigation and soil erosion

The dismissive view of Marakwet agricultural skills provided fertile ground for the febrile concern by colonial administrators about soil erosion and land degradation that was widely shared both in colonial Kenya and in the wider British colonial world (Anderson, 1984). The apparent environmental crisis and problems of overgrazing were acutely observed, and much discussed in the the Baringo lowlands to the east of the Kerio Valley (Little, 1992) and elsewhere in Kenya, notably in Machakos. There, the 'official mind' of the colonial administration linked 'environmental misuse' and 'the scourge of famine', while subsequent authors have mostly been quick to link environmental degradation and colonial land appropriation (Tiffen *et al.*, 1994). The power, and the flaws, of these 'development narratives' are alluded to by Tiffen (this volume).

In Elgeyo-Marakwet in 1944 it was reported that 'a considerable area of land which should be irrigable is now scoured by a large number of old furrows which have eroded to depths of up to ten feet in places.'[15] This perception of progressive degradation of resources in Marakwet did not fade through subsequent decades. The DC in Tambach commented on the Development Plan 1957–60 that it failed to suggest measures to tackle 'progressive deterioration of the hillsides overlooking Kibuswa, Talai, Sambirir, Mokorro Locations overlooking the Kerio Valley due to increased pressure of people moving into the Cherangani Hills'.[16]

It was suggested in 1944 that in order to stop gully erosion what was required was 'a system of correctly aligned and protected furrows giving on to land which should remain permanently cultivated with proper crop rotation'. It was recommended that there should be a vigorous anti-erosion campaign, that furrows not in use should be stopped, and that a

[13] R.H. Langridge to Carver 2/11/37; (DC/TAMB/2/2/16, 3).

[14] DC Tambach to PC Rift Valley Province 12/1/45, 'Dams: Construction and Maintenance of'; (DC/TAMB/2/2/16, 62).

[15] Report by P.V. Chance 19/12/44, quoted in Letter H. Hughes, Executive Engineer for the Commissioner for the African Land Use and Settlement Board to the PC Rift Valley Province, Nakuru 14//49 (DC/TAMB/1/11/11, 'Water', 61).

[16] DC Tambach to PC Rift Valley Province 24 May 1956, 'The Development Plan 1957–60; DC/TAMB/3/6/11, 'Development Estimates 1956–61, p. 6)

European should be appointed to execute anti-erosion measures. It was also noted that 'permanent occupation by one family of a given piece of land' was needed.[17] A number of central elements of colonial attitudes to Marakwet irrigation are encapsulated here: the need for settled permanent agriculture, for rational water use, for 'correct' furrow alignments (i.e. determined by engineering survey), and the need for European supervision. The irrigation of the Marakwet might save them from famine, but they were not managing it competently enough to avoid the need for intervention and social control.

Official views of the furrows seem to dither between two alternative views of Marakwet irrigators. On the one hand the irrigation furrows were impressive, and on the other they were seen to cause erosion, and irrigators to lack the skills and enterprise to manage them properly. Thus the DC wrote in January 1944 that 'the job of making new furrows should stand over until it can be done properly. The existing furrows are a fine piece of work and are essential for the peoples' food supply. At the same time, they appear to cause much soil wash and it would be inadvisable to extend their use until expert supervision is available.'[18] One solution to this logical dilemma was to stress the antiquity of the furrows, and the fact that current irrigators had not built them. Thus the DC at Tambach wrote in 1942 that 'the regulation of the water flow, the maintenance and construction of the furrows is all regulated by strict rules according to custom many years old.'[19] Interestingly, most authors writing about Marakwet irrigation repeat suggestions that the furrows are of ancient origin (e.g. Henning, 1941; Kipkorir, 1983; Moore, 1986). This view, at least in the colonial mind, appears to have allowed both a recognition of the value of what we would now call 'indigenous technical knowledge', and a view of contemporary lack of competence and zeal that justified (indeed demanded) European intervention and imposed change.

Irrigation and agricultural development

In 1943, the DC at Tambach wrote that 'The Elgeyo and Marakwet are shockingly backward tribes, and unless roused from their lethargy will soon become some sort of museum piece'; they hoped that the government would leave them alone, and he believed that they lacked 'the imagination to envisage any better sort of life'.[20] The people of the Kerio

---

[17] Report by P.V. Chance 19/12/44, quoted in Letter H. Hughes, Executive Engineer for the Commissioner for the African Land Use and Settlement Board to the PC Rift Valley Province, Nakuru 14//49 (DC/TAMB/1/11/11, 'Water', 61).

[18] DC Tambach to PC Rift Valley Province 14/1/44, 'Dams: Construction and Maintenance of'; (DC/TMB/2/2/16, 45).

[19] DC Tambach to PC Rift Valley Province 7/1/42 with 'Report of Dams: Construction and Maintenance of, 1/1/42'; (DC/TMB/2/2/16, 20).

[20] DC Tambach to PC Rift Valley Province 14/7/43 'Postwar Development and Employment'; (DC/TAM/2/1/54 Post War Development Plan Elgeyo-Marakwet District, 2).

Valley (despite the irrigation) were seen as 'the most backward part of the population'. The DC, meanwhile, saw unlimited possibilities for large-scale development in the 'potentially rich but dry country of the Kerio Valley'.[21]

The advent of the new imperative of 'post-war development planning' in memoranda from Nairobi led to a renewed emphasis on both the potential and the need for improvement in the irrigation systems of Elgeyo-Marakwet District. The effort and ingenuity involved in building the irrigation system could not be doubted ('people have already made gallant attempts to improve their water supplies by irrigation furrows'; in places 'the water was 'carried round the faces of precipices by birdsnest aqueducts which seem to defy the laws of gravity and hydro-dynamics'[22]), but the systems were 'primitive and almost certainly very wasteful of water'. Despite previous failure, it was confidently asserted that it would not be difficult to improve on the existing system: 'I have no doubt that technical advice, capital and labour could improve the water supplies out of all recognition'.[23] Extensive technical surveys were suggested, to include studies of potential dams and irrigation areas, boreholes and wells, tsetse control and experiments with new crops and agricultural techniques.

These wish-lists bore fruit in investment by the African Land Utilization and Settlement Board (later ALDEV) in Nairobi. Copies of plans of the Marakwet irrigation furrows were submitted to the Secretary of the African Land Utilisation and Settlement Board in Nairobi in 1948, and in 1949 the Executive Engineer for the Commissioner for the Board travelled along the Kerio Valley with the DC 'to determine whether and how the existing irrigation system could be enlarged and extended'.[24] Experimental intervention in the furrows began in 1959, with a view to possible extension for the next four years with the support of a tractor and a European supervisor.[25] Experiments with irrigated cash crops took place,[26] and provision for 'The Endo Scheme' were sought in the 1959 estimates.[27] The ALDEV estimates for

---

[21] DC Tambach to PC Rift Valley Province 14/7/43 'Postwar Development and Employment'; (DC/TAM/2/1/54 Post War Development Plan Elgeyo-Marakwet District, 2).

[22] DC Tambach to PC Rift Valley Province 14/7/43 'Postwar Development and Employment'; (DC/TAM/2/1/54 Post War Development Plan Elgeyo-Marakwet District, 2).

[23] DC Tambach to PC Rift Valley Province 14/7/43 'Postwar Development and Employment'; (DC/TAM/2/1/54 Post War Development Plan Elgeyo-Marakwet District, 2).

[24] Letter H. Hughes, Executive Engineer for the Commissioner for the African Land Use and Settlement Board to the PC Rift Valley Province, Nakuru 14//49 (DC/TAMB/1/11/11, 'Water', item 61).

[25] District Agricultural Officer to DC Elgeyo Marakwet 2 May 1959; (DC/TAMB/3/6/11, 'Development Estimates 1956–61', 61).

[26] DC Tambach note 'The Endo Scheme'; (DC/TAMB/2.2.12, 'Endo Irrigation and Furrow Scheme 1959–63', 1).

[27] DC Tambach note 'The Endo Scheme'; (DC/TAMB/2.2.12, 'Endo Irrigation and Furrow Scheme 1959–63', 1).

1960–3 for Elgeyo Marakwet District included £3000 for each of three years 'to extend the scope of the irrigation furrows from the principal eastward flowing rivers'.

The work proposed at Endo in 1959 was based on the water flowing in the Embobut River. It involved re-alignment of irrigation furrows on the floor of the valley using a tractor and disc plough. The furrows on the escarpment sides themselves were thought to be well-aligned, although in need of finance for improvement and maintenance.[28] The new (and 'correctly aligned') furrows on the valley floor would be put in and farmers charged at a fixed cost per acre, and 'controlled irrigation' from the primary furrows would be instituted according to a 'set programme' that would be 'agreed with the local people'.

By September 1959 the District Agricultural Officer was reporting that there was 'every indication that this scheme would be a success'.[29] By that time, 36 acres (about 20 holdings) had been ridged and furrowed and planted. However, technical confidence in the scheme was balanced by limited local support. Extension work and what was described as 'softening-up' had been done by the Assistant Agricultural Officer from Chebemiet, and had dealt with 'the 100 percent apathy' which hung over early *barazas*,[30] but there was 'a tremendous amount of opposition from the older age groups'. There was a fear among elders (who were seen as those controlling furrows) that water would run short if more land was taken into cultivation. New Instructors were being posted to the project 'to call together the more enlightened people and show them on individual holdings the advantages of this system of irrigation', and to carry out further work on 'the correct use of water and the introduction of cash crops'. Further opposition from 'the old men' was expected if the project ever expanded onto ridges not presently served by furrows, although by then it was hoped that 'the people may fully appreciate the benefits of the whole scheme'.[31]

Furthermore, the technical success of the new approach to irrigation was far from assured. Following a safari organized in October 1959 from Chesegon to Chechang, the Executive Engineer found that the enthusiasm of the project was not matched by technical competence. Fields had been badly levelled, and the work rushed such that it had needed to be repeated two or even three times. It seemed that neither the European Assistant Development Officer nor his staff had any engineering or survey training. The Executive Engineer acutely commented that 'without some basic knowledge of this subject you may

---

[28] A pipe or open channel section to carry water.

[29] DC Tambach note 'The Endo Scheme'; (DC/TAMB/2.2.12, 'Endo Irrigation and Furrow Scheme 1959–63', 3).

[30] Public meetings.

[31] District Agricultural Officer Elgeyo Marakwet to Assistant Director of Agriculture, Rift Valley Province 12 September 1959; (DC/TAMB/2.2.12, 'Endo Irrigation and Furrow Scheme 1959–63', 3).

have a very difficult time in your efforts to put the people's shambas to rights with the machinery that you have at your disposal'.[32] He wrote to the ADO, 'Since the Elgeyo [*sic*] method of irrigation is so very primitive and because the levellers that you have at your disposal do not know anything about it either, you must set yourself up as the expert on the spot and endeavour to secure the confidence and respect of the people without further outside help'.

This is not the place to catalogue the fate of the Endo Scheme, or its even less successful successors along the escarpment. Technical and socio-economic problems persisted, despite continued support from ALDEV, and the agronomic success with crops such as maize, chillies, and in controlling soil erosion.[33] However, this had been achieved at unreasonable economic cost. The DC wrote in May 1962 that he had obtained figures 'that make the Endo Scheme look very sorry for itself'.[34] The scheme had a substantial deficit, and there was no prospect of this being repaid. Either the scheme would have to pay its way, or close down and be sold up. Estimates suggested that the scheme could not cover its running cost in 1963, let alone repay the original loan for equipment. The DC concluded 'I must admit, I fail to see how this scheme ever got off the drawing board!'[35]

The ALDEV officer was withdrawn in 1962, and the scheme ran on under the agriculture department, with an annual subsidy from ALDEV for tractors. The DC justified running it in this way first 'because the scheme produces food in an area susceptible to famine', and second 'because it will (we hope) teach the Endo people how to irrigate without eroding the land'.[36] For the future, he urged a new focus on hand ridging, both because tractors were too costly, and because farmers' fields were too small for anything else. He also had a new crop to offer: bananas. These had been introduced only 10 years before, and grew prodigiously under irrigation: 'Banana plantations would help to rehabilitate Endo enormously, and if the people would abandon *wimbi* in favour of bananas the erosion problem would solve itself. A "Grow More Bananas" campaign should be tried.'[37] One senses a new panacea, perhaps a new 'received wisdom' in the making.

[32] Executive Engineer ALDEV to Assistant Development Officer Tot 22 October 1959; (DC/TAMB/2.2.12, 'Endo Irrigation and Furrow Scheme 1959–63', 10A).

[33] District Agricultural Officer to Assistant Director of Agriculture Rift Valley Province, Nakuru 22 May 1962 (DC/TAMB/2.2.12, 'Endo Irrigation and Furrow Scheme 1959–63', 195).

[34] DC Elgeyo-Marakwet to District Agricultural Officer Item 24 May 1962, (DC/TAMB/2.2.12, 'Endo Irrigation and Furrow Scheme 1959–63', 199).

[35] DC Elgeyo-Marakwet to District Agricultural Officer Item 24 May 1962, (DC/TAMB/2.2.12, 'Endo Irrigation and Furrow Scheme 1959–63', 199).

[36] DC Elgeyo Marakwet District Handing Over Note 2 January 1963 (DC/TAMB/2.2.12, 'Endo Irrigation and Furrow Scheme 1959–63', 244).

[37] DC Elgeyo Marakwet District Handing Over Note 2 January 1963 (DC/TAMB/2.2.12, 'Endo Irrigation and Furrow Scheme 1959–63', 244).

Discussion

In the event, it was agreed early in 1963 that all subsidy to the Endo Scheme should be ended, and that 'It was now definitely up to the irrigators to help themselves in this matter.'[38] They did so, of course. Over the last three decades all sign of the scheme has disappeared, except where it had left effective engineering improvements to the escarpment furrows. Bananas are now widespread along the escarpment, and despite the poor road access are among the most important cash crops. And irrigation continues, both at Tot and further South on other rivers down the escarpment. In the 1980s and 1990s there have been a series of new outside interventions by the Kerio Valley Development Authority (KVDA), and an NGO, World Vision International. There is also an important intervention in the valley by the District ASAL Development Programme. These projects have all in different ways sought to intervene in the Marakwet irrigation systems, both technically and socio-economically. They have experienced different degrees of success, small technical interventions being relatively successful and larger enterprises proving more tricky (Adams *et al.*, in prep. a). These projects have run up against identical problems to those experienced half a century earlier. Their organizers have been almost universally ignorant of their predecessors' failures.

The interesting thing about this fragment of a story about Marakwet irrigation is the way in which change on the ground has been driven by ideas drifting into official consciousness in Nairobi, and percolating down through the bureaucratic system until they come to rest in the in-tray of someone who has to make something happen, or at least appear to make something happen. Received wisdom about environment and development is mediated through this bureaucratic process, and emerges in mixed and sometimes unexpected forms in practice. Discourse within the community of development 'experts' is itself complex in space and time, and shifting in content. There are many layers to this discourse, many barriers to communication, much creolization of thought, language and perception. What eventually happens on the ground is indeed often greatly influenced by the ebb and flow of ideas from outside, but it is also highly opportunistic.

Paralleling the diversity of this process is a remarkable evolutionary continuity in what actually happens on the ground. New ideas do not necessarily rush in through some kind of Kuhnian 'paradigm shift', ousting all predecessors. They may do so in the fiercely argued world of academic debate, but out in the bush they seem to succeed each other in a different and much softer and more complex way, sedimenting down gradually and adapting themselves to preceding thinking. In Marakwet old ideas have an astonishing continuity. Food self-sufficiency, soil

---

[38] Minute Provincial Agricultural Committee E/14/63 (DC/TAMB/2.2.12, 'Endo Irrigation and Furrow Scheme 1959–63', 266).

erosion and overuse of land, and the need for 'modern' methods are still central themes in the way outsiders are thinking about the future for Marakwet, and underlie their various projects and initiatives. Development narratives may appear to come and go, but sometimes they are but new bottles for old wine as, indeed, Roe (1991) suggests they should be.

Development narratives are the means through which ignorant but enthusiastic outsiders make sense of complexity; the source of their confidence (often of course misplaced) that they can diagnose problems and prescribe useful solutions; and (of course) the standards by which short-term 'success' and 'failure', and hence career prospects, are judged. The impacts of the formulaic thinking of outsiders in the development process can be considerable (and locally dramatic and sometimes catastrophic). The impacts of received wisdom, and changes in received wisdom, on local communities (i.e. the historical pattern of 'development' in particular places) deserves extensive research. However, this must be balanced by research on the outside development agents themselves. Perhaps even more importantly we need to study the institutions where they are trained, employed, assessed and promoted, and from which they issue forth on their mission of cultural, social, economic and environmental transformation.

10

Land & Capital

MARY TIFFEN

Blind Spots in the Study of the 'Resource-Poor' Farmer

Introduction

Academics and civil servants need neither capital nor land, nor do they depend on profits for family survival. In developing countries, consultants are rarely even paid and briefed by farmers, as may happen in Europe and the United States. Perhaps because of this, the significance of the land and family capital at the centre of both the affections and the efficiency of family farms may not be recognized. A manual on the methodology of irrigation feasibility studies in Europe, for example, makes calculation of farm incomes central, together with an emphasis on the finances of the management authority (Bergmann and Boussard, 1976), while guidelines produced by the World Bank and the UN Food and Agriculture Organization (FAO) relegated both of these considerations to a much less important place (Tiffen, 1987). Agricultural economists in Europe must take full account of the rights and income streams of their clients, the farmers; while most studies of developing country agriculture are carried out by and for salaried employees in universities, governments, aid agencies and non-governmental organizations (NGOs).

This leads to a perhaps unconscious predilection in agricultural development in favour of external interventions which expand, or at least maintain, the role of the public sector. Expatriate jobs and consultancies depend on the size of aid budgets. External intervention will be handled by national civil servants, whose interest is in securing jobs and resources for their organization. It suits to view farmers as destructive of the environment (due to poverty, ignorance, or greed) and requiring direction, or at least, advice. An increase in government control over land and water resources can also increase political control, and may also bring financial perks or reduce government costs.

Even when intervention is channelled through externally funded NGOs, aid agencies benefit, gaining support from those who might otherwise be among their critics. It appeals to public opinion, which sees young Westerners actively doing good to the poor. It appeals to idealists, who would like to find among the poor laudable communal arrangements lost in industrialized societies. Great efforts are made to find and describe what may be the remnants of such organizations (which will have to be helped), or a short-term success of one NGO in one or two villages (which should be replicated). Both will create jobs for interventionists. The impact on the welfare of traders who have stimulated the growing of a new and profitable crop is ignored because it fits neither the utopian vision nor the covert interests of the intervenors.

A related predilection among academics is for researching subjects that are fundable and publishable (currently, environment, gender and participation), leading to recommendations generally expensive in government resources and usually at odds with standard practice in industrialized economies today, let alone when they were poorer.

It is important to be aware of these predilections on the part of stakeholders in development, since 'blind spots' or illusions about land tenure and capital lead to a failure to recognize: first, the tendency for tenure to shift towards more individual rights as population density increases, and small farmers' intensely possessive regard for their land once it has become scarce; and second, the way in which even poor people can find capital for what is really profitable, and the importance of that capital in raising the productivity of agriculture as land becomes more scarce.

In relation to land, there is a tendency to romanticize common-property arrangements without serious effort to find out whether they are locally preferred by potential users and non-users, whether they complement individual land rights which may have already become established, or whether they are feasible within the local socio-political reality. Communal tenure has administrative attractions, because it seems easier to control a few 'traditional' authorities than many individual owners (as evidenced in the former homelands of South Africa), and it may seem to facilitate anti-erosion measures and environmental improvements.

The conventional reaction to a perceived capital constraint facing a project deemed socially, politically, or environmentally desirable is to introduce a government- or NGO-managed credit scheme, rather than to investigate why capital is not already flowing in that direction. The sources of capital, and many of its functions, are frequently seriously under-estimated, along with its gradual accretion over time.

This chapter will illustrate these biases in agricultural development research and practice, mainly with reference to Machakos District, Kenya. Here, the positive effects of certain external interventions on farmers' willingness and ability to invest in productivity improvements, and the hindrances caused to farmers by the 'blind spots' identified above, were studied over the 60 year period 1930–90.[1] A serious examination of the process of change revealed substantial improvements in both farmers' welfare and land productivity, in an area where short-term observers could see only degradation.

Land tenure

Ester Boserup noted the way tenure and land use change as population densities increase:

> For instance, long fallow cultivators have a right to clear land in areas belonging to their community. When long-fallow agriculture is replaced by . . . permanent fields, it is necessary to change to another system of tenure, which gives the cultivator more permanent rights to the land he is cultivating. Under land use systems based on the grazing of fallow land, some of the cultivators, or some non-cultivating members of the village community, enjoy the right to graze animals on the fallow area for a certain part of the year . . . The substitution of a more intensive system for the old one requires the abolition of the grazing rights. (Boserup, 1970: 109)

Colonial perceptions of land rights and environmental degradation
The Akamba of Machakos District had concentrated their settlements on the better watered and defensible hills, and this led to higher population densities than in many parts of Africa – already over 80/km² in the hills in 1930. Private rights over land were evident in 1910–12, when an anthropologist observed that 'Each man owns the land he and his family cultivates . . . On removing from the place one of the family is left behind to look after the fields, or they are left in the care of some relative or neighbour. Or else they are sold for one or two goats . . .' (Lindblom, 1920: 165).

Land in the lower hills and plains was known as *weu*, unpopulated and unappropriated (but not common property, which implies that it is managed by a community institution). People could establish an exclusive grazing area (*kisese*) in the *weu*, around a cattle post, by marking trees. Their rights continued so long as they used the area. If they left it, it reverted to open access. If young men, herding together, decided to establish a new village, they cleared land for their wives to cultivate, which then became their property in perpetuity, even when fallowed. Thus, *weu* could be converted into private land by cultivation. Once private, the owner could allow another to establish a *kisese* on it, but he would be a tenant, who would have to quit if asked.[2]

[1] This paper draws on work done in a collaborative study with the University of Nairobi, which has been summarized in Tiffen *et al.* (1994).
[2] The Reconditioning Committee in 1936 established this when it asked its two native members about tenure. DC/MKS/12/2/2 - Minutes. Kenya National Archives, Nairobi.

Pressure on the *weu* in the most densely inhabited areas led to disputes between older, wealthy cattle owners, who wanted to use the remaining areas for grazing, and younger, poorer men, who wanted to use them for new farms. The Provincial Commissioner intervened in 1927, in an area near Machakos town, to uphold first-cultivation rights, provided the plough was not used. Land shortage was aggravated by the small but growing numbers of plough owners in areas close to a market, who were enlarging cultivated areas to make sales. As the *weu* diminished, people made more permanent use of their *kisese*. Some began to put thorn hedges around it, the better to defend it from other people's cattle. This put more pressure on the remaining nearby *weu*, as a source of grazing, building timber and fuel (Munro, 1975), leading to a noticeable increase in soil erosion on grazing lands deprived of vegetation and on cultivated lands unskilfully ploughed.[3]

The government wished to combat soil erosion, which had almost as high a political profile in the 1930s as the environment does today. It became inconvenient to recognize the Akamba as full owners of their land, although in fact administrative files and reports in the 1930s refer to owners, and to sales of land. Technical experts wanted to move cultivators off the steeper slopes (i.e. evict them, temporarily or permanently, from their land), and to use tractors for terracing.[4] Although the Akamba had participated in early anti-erosion experiments, they now began to fear losing more land to European settlement, especially after another anti-erosion measure was carried out: the forced sale of their 'surplus' cattle in 1938.

Administrators were made cautious when the Akamba secured a reversal of the cattle seizures through a sit-down protest in Nairobi and lobbying at Westminster. In a confidential memo in September 1938, written after discussion with the agricultural and soil conservation officers, the District Commissioner, G.J.L. Burton, agreed that the state must take final responsibility for the care of land and, therefore, ensure that all farms were sufficiently large to allow a proper proportion of pasture to arable land, to prevent over-grazing and erosion. However, the division of the reserve into farming units of economic size should be done, somehow, by propaganda and persuasion, achieving consolidation through sales and exchange. To maintain the ideal size, land was to be inherited by one son only,[5] while Akamba custom was – and still is – inheritance by all sons of the mother who farmed the land.

Most administrators and agriculturalists preferred to believe that African land was managed by tribal authorities and that individuals

---

[3] The reality and extent of the problem is shown by photographs taken in 1937. Some of these are reproduced in Tiffen *et al.* (1994), accompanied by photographs of the conserved landscape in 1990–91.

[4] Barnes, R.O., 1937, Soil Erosion, Ukamba Reserve. Report to the Department of Agriculture Memorandum, July. DC/MKS/10a/29/1, Kenya National Archives, Nairobi.

[5] Joyce: MSS Afr s 2157/8, Box 5. Rhodes House Library, Oxford.

held only use rights. This conveniently implied that a right to use another piece of land could be substituted if the government for any reason wanted to dispossess the user. White settlers, interested in obtaining more land, supported this view, as well as the belief that Africans were destructive cultivators (Anderson, 1984). Permanent individual rights were seen as something new, and to be opposed. The new Governor in December 1944, Sir Philip Mitchell, pinned his reputation on enforcing a successful anti-erosion campaign, using the concept of a tribal land authority, and prioritizing Machakos (Throup, 1987). The District Commissioner, Brown, more closely in touch with Akamba custom, expressed his worry in January 1945 to the Provincial Commissioner:

> Arising from His Excellency's decision to press forward the Machakos post-war reconditioning plan . . . reassurance will be necessary on the land question . . . In applying His Excellency's policy of regarding all native land holders as tenants of the tribe, I believe that the sanction to enforce proper cultivation should be that the Land Authority if need be does the necessary work recovering the cost from the land holder, if need be by distraint on the crops. We should I think not say too much about evictions of bad tenants . . . We should say we are respecting existing rights even when the land is closed and evacuated for reconditioning . . .[6]

In April, Brown wrote:

> Merest hint of a readjustment of holdings in *itui* has produced widespread nervousness and a fear that the government is about to take away their land. When the Works Company went to Momandu to camp, in preparation [for some rehabilitation work], it was met by flat opposition as they believed it a threat to their land rights. I was told they did not understand what this company meant and would prefer to do the work themselves.[7]

He further said that if the planned works were not stopped, he must ask to be relieved of responsibility for them.

Meanwhile, Grieves, the Agricultural Officer in Machakos from 1938 to 1946, being one of several agriculturalists who admired the Russian solution to the perceived inefficiencies of small holdings, planned for supervised, collectively owned farms at the proposed new settlement area at Makueni. Individual title was granted only after adamant opposition from the local Council, which included a stronger elected element after 1946.[8]

The Government was sufficiently worried about the unrest to send H.E. Lambert, a Senior District Commissioner and their expert on

---

[6] Joyce: MSS Afr s 2157/8, Box 5. Rhodes House Library, Oxford.
[7] Joyce: MSS Afr s 2157/8, Box 5. Rhodes House Library, Oxford.
[8] Makueni was eventually settled on the basis of individual title, and the Akamba have ignored the original condition which required them to nominate only one heir.

African tenure, to investigate. Lambert favoured the communal view of African tenure (Throup, 1987), but was critical of the detail of Grieve's Makueni plan. Although he had to recognize the strong individuality of existing tenure, including customary sales, his predilection shows in his proposed land authority, which was to consist of leaders from the main clans, under strong European supervision, and with well-trained staff, demonstrating an emphasis on salaried jobs and on control. This authority would compel 'kinships' with unutilized land to accommodate those from overcrowded areas, by lending or renting (preferably the former), so as to provide as many people as possible with an economic holding. This ignored both his own statements that the *itui* or village, which was not necessarily kinship based, was more important than the clan; and that it was unlikely that those with big farms would happily cede land to the poor. His intellectual convolutions in support of a policy he thought would lead to less poverty and less erosion are now obvious, although there may be sympathy for his motives:

> There remains the question as to the degree of recognition to be accorded to the native idea of actual ownership of land. We need not interfere with ownership directly; nor need we put a stop to sales . . . However this principle [beneficial occupation, by which he meant the equalization of land holdings referred to above] will mean that many families are occupying land that is owned by other families.[9]

He hoped that

> Gradually this emphasis on use as opposed to ownership might be expected to enhance the 'traditional conception that the community has an ultimate concern in the control of the use of land' [quoting Lord Hailey's report],

and that 'Sales and the need for registration of land transactions may slowly fade away'.[10] It would appear that such a land authority was found impractical. The Akamba conception of ownership was with the family, not the clan, and there had never been clan authorities deemed to own land.

In October 1945 the Machakos Reconditioning Committee, on which Akamba representation had been reduced to one government-appointed chief, approved plans to recondition some areas with tractors to speed up the work.[11] To the astonishment of the Committee, the Akamba responded to this 'help' by lying in the path of the tractors (Peberdy, 1958). In the Akamba view, cultivation gave property rights, and the

[9] Lambert, H.E., 1945, A note on native land problems in the Machakos district with particular reference to reconditioning. DC/MKS/7/1, 140–4. Kenya National Archives, Nairobi.
[10] Lambert, H.E., 1945, A note on native land problems in the Machakos district with particular reference to reconditioning. DC/MKS/7/1, 140–4. Kenya National Archives, Nairobi.
[11] DC/MKS/12/2/1. Kenya National Archives, Nairobi.

tractors illustrated the government's claim over their land. To the regret, sometimes loudly expressed, of settlers and some agriculturalists, the District Commissioners in Machakos succeeded in proceeding on the basis of consent (on the land issue) and compulsion (on the work issue). A touring report on one area in early 1948 notes that the steep slopes of the hill had been closed to grazing with the consent of the *kisese* owners concerned. Some young men had been prosecuted and fined for failing to turn up for the two days a week of compulsory work.[12]

The DC's 1948 Annual Report still blamed 'the idea of individual tenure' as opposed to 'clan tenure' for slow progress in correctly laying out terraces and drains, and the difficulty in getting land set aside for forestry, although it recognized that individual tenure had encouraged care of the soil by progressive farmers. However, a new agricultural officer, Hughes Rice, arriving in 1951, and the DC in 1952–3, Penwill, whose exposition of Akamba law recognized the validity of their customs on ownership and inheritance (Penwill, 1951), worked together in instilling confidence that land rights would be respected. Hughes Rice was not academically trained; he had joined government service in 1938 after 10 years in private farming. In an earlier posting at Fort Hall he had realized that group farming was not workable due to the deeply engrained desire to possess land, shared by Kenyans in common with other peasant societies. He also realized the necessity of recommending profitable activities, in addition to straight soil conservation, and the vital importance of winning the support of people by securing their understanding of, and consent to, proposed measures.[13] Within the Reserve area, there was an improvement in cooperation between government and people around 1952–8, particularly when people were given more control over the methods of work and choice of technology,[14] although this broke down in the run up to independence.

Conflict between government views of land rights and those of the Akamba remained, however, over the Yatta Plateau. The Akamba regarded this land as *weu*, available to those who chose to make new farms. The Government regarded it as unoccupied Crown Land, where it could specify land use and regulate, or forbid, access. It was managed by the Veterinary Department, which by 1950 had delineated five blocks for rotational, licensed, communal grazing. Settlement and arable use were forbidden. The aim was 'the conversion of a piece of raw Africa into a well-managed grazing area in which the number of stock is related to the carrying capacity of the land' (Government of Kenya,

---

[12] DC/MKS/8/5. Kenya National Archives, Nairobi.
[13] Hughes Rice: Mss Afr s 1717, 84a. Rhodes House Library.
[14] Farmers came to prefer the bench terrace over the initially compulsory narrow-based terrace, although it required more labour. The organization of work was gradually shifted from village mobilization under the direction of chiefs (government officials) to self-help *mwethya* groups who chose their own leaders and technology (fuller details are in Tiffen *et al.* 1994).

1962: 277). Despite, or because of controls, a survey in 1957 found the Yatta Plateau suffering bush encroachment, tsetse infestation and, on areas deemed overgrazed, the dominance of undesirable grass species.

In the run up to independence in 1963 the land-use controls collapsed and the Akamba swarmed in, burning and slashing. Their destructive 'shifting agriculture' techniques were condemned even by sympathetic observers (e.g. Mbithi and Barnes, 1975) who did not take time to ask their reasons. In fact, rotational cultivation round a settled base gave the pioneers tenurial rights respected by other Akamba, on the basis of which, over 20 or so years, they have invested in soil and water conservation structures on their arable land, planted and protected trees, and allowed part to revert to bush grazing, private because once cultivated.

Some livestock specialists think communal grazing necessary to give flexibility to livestock owners in times of drought (Behnke and Scoones, 1993; cf. Brockington and Homewood, this volume). This clearly depends on particular site conditions, however. In Machakos, small mixed farmers obtained the necessary flexibility in the 1984 drought by supplementing grazing with crop residues, by renting in land, or by buying in fodder (ADEC, 1986); they lost proportionately fewer animals in the arid south than neighbouring Maasai on large group ranches (Mukhebi *et al.*, 1991).

Land ownership and registration of title
It is conventionally thought that registration of land title is necessary to give the security of tenure that promotes investment. The experience in Machakos shows security of tenure is important. Formal registration by legal means may enhance this, but is not absolutely necessary where customary tenure already provides sufficient protection. The Swynnerton reforms of 1954 represented a victory within the Agricultural Department for those who favoured individual land title registration and land consolidation, first in Kikuyu areas and later in areas where African farmers were resettled on land purchased from departing Europeans. Consolidation of small and scattered fragments, which were regarded as wasteful of labour and unconducive to a good farm plan, was initially a condition of registration. As the Akamba valued having land in different ecological niches, they refused titling until the condition was dropped. Registration of their land only began in 1968. Even by 1992, registration in all its stages had been completed in only 35 per cent of sub-locations (Tiffen *et al.*, 1994: 66). It is customary arrangements which have given sufficient security for investment – and for sales. Some 28 per cent of farmers in a densely populated area had in 1964 acquired some of their land by purchase, although the percentage was smaller in less-densely settled areas. A survey in four locations in 1980 found that in the previous five years, 21 per cent of farmers had bought some land and 15 per cent had sold (Meyers, 1981).

Effects of blind spots on tenure

Reluctance to accept the capacity of customary tenure arrangements to evolve dynamically, and the convenience of 'communal' control for the state, has also been manifested in other parts of Africa. In Nigeria, central administrators and academics continued to state that a farmer was prohibited from transferring the land of his or her community to 'strangers' (Oluwasanmi, 1966:193) long after sales had become commonplace in some areas. For example, when a new railway was built in 1956 through Gombe Emirate, a northern area which had received rural in-migrants from distant areas, higher compensation for land was paid than in less-populated areas, because it was recognized that farmers could obtain alternative land only by purchase. The Senior Councillor of the local authority formally recognized some land sales in 1960, agreeing the right of people to sell land they had themselves cleared from bush. By 1970 the local judge recognized also the right to sell land which was inherited, or bought. By the time of interviews in Gombe in 1967, around a half of those who had enlarged their farms had bought at least some of the extra land (Tiffen, 1976). However, the mainstream academic and official view became enshrined in the Land Use Decree of 1978, reducing the possessor's interest to a right of occupancy. This was utilized to pay unrealistically low compensation to farmers losing land to irrigation schemes in Sokoto and Kano, causing much hardship since there also tenure arrangements had evolved to something close to freehold, sales were common, and land was practically unavailable by other means (Wallace, 1980; Bird, 1983).

Although sales have become commonplace in parts of sub-Saharan Africa, almost no academic work is done on land markets. By contrast, villagers, if asked, have no difficulty in giving the average price of land, and discussing the way this differs according to improvements made on it or to its location. Farmers in a coffee area of Machakos valued land planted with trees in 1990 at Ksh 80,000 per acre, and land without trees at Ksh 40,000. Women in Kabale District, Uganda (though disclaiming detailed knowledge because men owned land) were in agreement in pricing a road-side field at three times a similar sized one half way up a hill.[15]

The right of sale is often viewed askance by social scientists, who assume that it will lead to impoverishment. In many circumstances, it should rather be seen as giving people added flexibility in response to new opportunities. People who had very small farms in the highlands of Machakos in the 1960s and 1970s were among those who sold up to finance a move to the more arid lowlands, where they could clear a larger farm. Most felt they had done better by the move (Matingu, 1974). In parts of Arua District, Uganda, where clan control still prevents sales, people are unable to raise capital for a non-farm business or a new farm elsewhere by this means.[16]

[15] Personal field observations, 1994.
[16] Personal field observations, 1994.

Capital

It tends to be expected that incomes decline as farm size declines. Since reduced farm size is an inevitable consequence of population growth when the inheritance system divides the asset amongst several heirs, it is commonly assumed that population growth must increase poverty. However, capital can to some degree substitute for the factor of land. Income can vary, therefore, independently of farm size.

Population growth and associated market forces drive changes in farming systems. Almost all observed systems were different 20 years previously, and may be unrecognizable in 20 years' time, as new technologies and management innovations are adopted, and as the ratios of capital and labour to land change. A farming system is often part of an income-earning strategy that encompasses both farm and non-farm activities. Farm studies are defective if they neglect the factor of capital, the dynamics of change over time, and family and other linkages with the non-farm economy, which often provide access to capital.

Upton's standard textbook, *African Farm Management*, notes that capital includes not only machines and tools, but also buildings, roads, footpaths, drainage ditches, terraces, irrigation equipment, growing crops, livestock and stocks of food, seed, fertilizer and other materials (Upton, 1987: 8). With the exception of roads, these are mostly privately obtained. Figure 10.1 gives a diagrammatic representation of a two-hectare farm belonging to a man who started farming in Machakos in 1970, following employment as a waiter in Nairobi – scarcely a high-paid occupation. By 1990 he had improved his farm with cut-off drains diverting roadside run-off for crop use, bench terraces, grafted fruit trees and pits for bananas, live fencing as a windbreak, and water storage in a pond for vegetable cultivation (Mortimore *et al.*, 1993).

Most agricultural surveys in developing countries neglect such items of fixed capital. As current received wisdom describes those under survey as 'resource-poor' farmers, it seems unnecessary to bother about their resources. However, as Hagen said long ago, 'We could not save on their incomes, but they can' (Hagen, 1964: 18). Upton (1987) cites studies in Nigeria and Zambia that show that cash-cropping farmers may save as much as one-third of their incomes.

Ester Boserup thought 'Nobody will object to the statement that the feeding of an increasing population from a given area requires an increase in the amount of capital' – the increase per head of agricultural worker being necessary so as to compensate for the declining area of land at his (sic) disposal (Boserup, 1970: 105). She took the increase in capital for granted, for received wisdom in the 1960s represented development as being mainly the result of investment. In her seminal study (Boserup, 1965) she therefore concentrated on the fact that agricultural intensification also required increased inputs of agricultural labour.

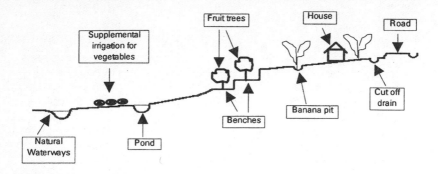

Figure 10.1   Diagrammatic representation of a Machakos farm. From drawing by
F.N. Gichuki.

Capital in peasant agriculture has since become an unfashionable
subject. Of 20 different farm studies made in Machakos District between
1963 and 1986, for instance, only the first considered capital availability
(Heyer, 1966), although one concentrated on credit (Meyers, 1982).
Heyer found that capital was available for investment in ranches and
transport enterprises, but not for the then unprofitable agriculture. Most
other reports only considered capital in the form of an ox-plough. One
included buildings, but it did not appear to be very accurate. Not one
study gave information on costs of trees planted, terraces, dams, ponds
or other constructed water sources, or fencing. Yet farm incomes in
Machakos depend crucially on whether the farmer or his or her
predecessor has invested in these elements, as can be illustrated with
reference to terraces.

Terraces
Attempts in Machakos to compare yields and incomes from terraced
and the (few) unterraced fields were frustrated by extreme weather in
the seasons of observation (Figueirredo, 1986; Holmgren and Johansson,
1987), although the findings implied that terraces gave substantial
benefits in the more humid areas when the rainfall was below average,
and in all seasons in the drier areas. The benefits came from two
sources: higher yields from the main food crops, because of better water
retention; and income from fruit and vegetables which could not
otherwise be grown. The best study comes from the neighbouring
district of Kitui, which had more unterraced land than Machakos,
making it easier to find matching pairs (Holmberg, 1990). Table 10.1
shows the high rewards to terrace construction, after taking account of
the additional labour requirements for terrace maintenance, and the
opportunity cost of the capital.

Table 10.1   Economic value of soil conservation, (Ksh/farm)

	Farm without soil conservation	Average soil conservation practices	Superior soil conservation practices
Total revenue	7,550	12,187	14,732
Variable costs			
Food crops	853	853	853
Soil conservation crops*	–	170	230
Total variable costs	853	1,023	1,083
Gross margin	6,697	11,164	13,649
Annualized costs			
Investment in soil conservation	–	165	165
Gross margin minus annualized costs (net return)	6,697	10,999	13,484
Labour requirement (person-days)			
Cultivation	585	636	692
Soil conservation work	–	45	45
Total labour requirement	585	681	737
Net return for person-day of labour	11.44	16.15	18.29

Source: Holmberg (1990: 66)
*Crops feasible only with soil conservation

Heyer (1966) noted the terrace as an investment but did not calculate the return to it. Instead, she took the repair status of terraces as one indicator of good management, and calculated returns to management. She found that good managers made three to four times as much as poor managers.

In Machakos District some terraces have been constructed by rotational work groups (*mwethya*) which move from farm to farm, and others by hired labour. *Mwethya* can be regarded as the manufacture of capital through labour: the participants sacrifice leisure by working on other people's farms in order to get the necessary input of person-days on their own farm. However, most of the farmers we interviewed (whom we do not claim to be a representative sample), said their terraces were constructed by hiring labour, using capital from a variety of sources, including non-farm earnings, livestock sales, and farm profits. No-one has investigated the proportion of terraces constructed by hired labour. Academics and journalists have concentrated on the *mwethya* groups, attractive within current development fashions because 'participatory' and often with female majorities.

Trees

The second most important form of fixed capital is trees. Trees in Machakos have considerable establishment costs, since coffee, citrus and some others have to be planted in pits or cut-offs, and provided with a heavy initial dose of manure to improve water permeability as well as to give nutrients. The substantial quantities of manure required may be purchased by the lorry load from nearby ranches. The planting material may also have to be purchased. Tree nurseries are run not only by the state, but also by large export-orientated firms, enterprising individuals and by the women's groups, the latter attracting most academic interest. As we have seen, tree-planting costs are reflected in land prices.

Unfortunately, there were few farm management studies relating to the coffee areas. In the semi-arid area, one study (Hayes, 1986), while noting a substantial increase in the planting of fruit trees, concentrated only on the effects on fuel supply, which is the fashionably perceived use, but not the only use, of trees. Farmers derive income from trees in the form of fruit for own use and sale, own use and sale of firewood and construction materials, sale of industrial inputs (tanning extracts, fuel for brick making) and they may also form inputs to other farm enterprises. Banana trees, for example, are used as livestock feed after fruiting. Coffee in the higher areas and citrus, pawpaws and mangoes in the drier areas are now important earners, due to efficient water management techniques. Eucalyptus or wattle are grown on steep, otherwise uncultivable slopes, and coppiced for a timber income. Some farmers grow fruit on a recognizably commercial scale, with from 10 to more than 100 trees; others have less than 10 trees. The latter may be ignored in a farm management study, but they can be important either as an income source, or in providing working capital for the main crop.

Coffee illustrates the capacity of farmers to invest when the right incentives are present. The boom in coffee prices in the late 1970s led to a surge in coffee planting in Machakos, which was recorded, since coffee has to be sold through cooperatives to the Coffee Board. Each additional hectare put down to coffee represents investment in terracing (without which coffee cannot be grown in Machakos), in the purchase of manure for the planting pits, the purchase of nursery stock, and most probably labour hire. The land investment shown in Figure 10.2, which was replicated in other coffee-growing areas, and duplicated by farmers planting oranges and pawpaws in the semi-arid areas, was not included in official statistics of national capital expenditures. What is not measurable through official statistics is frequently ignored as a factor in development, so we continue to talk about resource-poor farmers and low-input farming.[17]

---

[17] Farming systems with low inputs of capital and labour are feasible and often preferred where land is plentiful; in these systems it is the land input that is high.

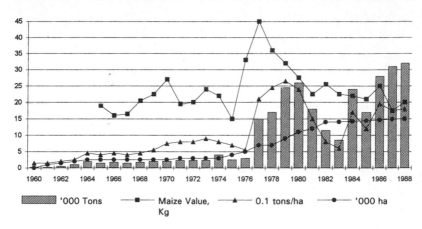

Figure 10.2   Coffee areas, yield and output, 1960–90, Machakos District.
Note: Value is calculated in terms of purchasing power for maize.

Neglect of the value and ownership of trees can be as convenient to government and project designers as neglect of land rights can be, with almost as devastating consequences for farmers' income. In one section of a government irrigation scheme for small farmers in Zimbabwe farmers earned Zim$95 from mangoes in the year of observation, while those in a neighbouring section averaged Zim$1. The second block had suffered a government 'rehabilitation', during which almost all trees had been cut down without consultation or compensation (Tiffen, 1990: Table 2.1).

Buildings and stocks
A third important form of capital is buildings and stocks. In the variable climate of Machakos, a well-built, well-filled granary (usually built with purchased timber) obviates or delays the necessity to sell farm assets such as livestock to buy food in bad seasons. Sale of such assets depresses future incomes. Again, this can best be demonstrated by reference to the Zimbabwe irrigation study. Two of the four blocks were particularly vulnerable to water shortage when the main river feeding the scheme ran low during drought years. On the water-risky blocks fewer livestock were owned because of past sales, reducing present livestock income. Farmers had less equipment including ploughs, since they were less able to save. This depressed net crop income, through higher costs for hiring ploughs and late planting. More plots were managed by women, since many men had migrated out in search of paid work (Tiffen, 1990: 14–16). The appropriate response to this situation is not necessarily to provide credit or to pay special attention to women farmers, since both men and women farmers on the two better blocks

had demonstrated the ability to save for profitable equipment. The solution was more likely to lie in the improvement of water supply – the enabling factor in this particular situation – without which credit would simply lead to debt.

Other important capital investments evident in Machakos are in fences or hedges to contain livestock, in pens where they are kraaled at night (and in some cases, permanently kept under a zero-grazing system), and in the roofs and guttering that trap water, thus saving women's labour for more productive activities than water collection.

Working capital
Working capital is also important. One Machakos farm study classified expenditure by farm size quartile. The researchers (Rukandema *et al.*, 1981) did not comment on the implications of the capital findings, which are brought out by some additional calculations (Tiffen *et al.*, 1994), as shown in Table 10.2.

Contrary to expectation, total farm income was bigger on small farms than on medium-sized farms, being only a little less than that of the largest farms. Cash inputs other than labour were highest per hectare on the smallest farms, giving them by far the highest net farm income per hectare. The smallest also had the second highest cash expenses for labour, though much less than the largest farmers. Even small farmers

Table 10.2   Farm strategies on large and small farms in Mwala, Machakos District, 1980

Quartile:	1	2	3	4
Farm size (ha)	1.3	3.24	7.54	17.8
Cropped area (ha)	1.02	1.62	1.92	3.24
Cattle owned	4	5	8	11
Sheep/goats owned	7	12	16	16
Cash expenses (Ksh):				
labour	287	164	23	1007
other inputs	338	309	320	438
Cash inputs/cultivated ha	331	191	167	135
Return/cultivated ha to inputs	4.89	3.19	1.6	1.14
Total net farm income	2,105	1,980	2,050	2,736
Net cash farm income	636	464	803	605
Subsistence farm income	1,469	1,516	1,248	2,131
% subsistence in farm income	70	77	61	78
% farm cash from livestock	57	60	78	66
Net farm income/ha	1,619	611	272	154
Off-farm income	3,529	1,503	1,811	2,628
Per capita income*	626	387	429	596

*Assumes 9 persons per family in each class. No data given to show if family size varied by farm class.
Source: Rukandema *et al.* (1981), and own calculations.

find it advantageous to use working capital to supplement family labour to secure timely operations.

Table 10.2 shows that the smallest landholders had, on average, a fifth of the land area (1.3 ha including 1 ha under crops) and half the average number of livestock (4 cattle and 7 sheep or goats) owned by farmers in the third quartile. Rukandema *et al.* do not tell us the purpose of the cash expenditures; they may have included fodder. More than half of small landholders' farm income came from livestock.

It is difficult in a one-off farm survey to expect the farmer to recollect accurately small sales of fruit, vegetables and poultry made at frequent intervals. It is easier to ignore them, however much they may add up to in total. Table 10.3 suggests this might be dangerous. It gives the percentage of households in Machakos who said they got income from various sources, some commonly neglected. It is not comprehensive: ADEC (1986) neglected to ask about non-farm businesses, building poles, and fruit and vegetables (which farmers may or may not have included when asked about cash or food crops).

It is equally tempting to avoid facing the difficulties of calculating livestock income accurately. Livestock are the original form of capital; they frequently have to be purchased in, for example to replace plough oxen lost in a drought, or to up-grade stock. Income should theoretically include gains and losses from births, deaths, sales and purchases, distinguishing between revenue and draw-down or additions to capital; and sales and consumption, net of costs, of livestock by-products such as milk. There are difficulties in the treatment of manure and draft, both outputs of the livestock enterprise which are used in the crop

Table 10.3   Household Income Sources, Machakos District (% households mentioning)

Agro-ecological zone:	Semi-humid	Better semi-arid	Poorer semi-arid
Wage/salary	48	41	46
Cash crops	74	53	53
Food crops	38	48	58
Cattle	28	52	56
Goats	29	56	65
Milk	19	15	14
Chickens	38	53	49
Eggs	20	21	19
Hides/skins	21	34	40
Crafts	36	43	36
Pombe (beer)	1	2	3
Honey	0	9	18
Charcoal	7	11	14
Firewood	9	5	5

Source: Calculated from ADEC (1986), Table 3.10

enterprise; and of crop residues, an output of the crop enterprise which can be used in the livestock enterprise, or used as fuel, or as a mulch. All these things can be, and sometimes are, sold for cash.

The sources of capital

Almost the only certainty about the sources of farm capital in Machakos is that formal credit played an extremely small role, except in the case of working capital for coffee, where part came through coffee cooperatives. In other studies in Kenya it has been found that inter-family transfers are important. Non-farm incomes (often, but not exclusively, of men) generate savings that are sent back for wives to invest and manage on the farm. Alternatively, the returning migrant brings home savings, and invests in a non-farm business such as a shop which then generates capital for the farm. Children, for whose education much tends to be sacrificed, are expected to provide gifts, some of which are for investment purposes.

Non-farm incomes often provide the kick-start needed by an impoverished agriculture. Livestock sales are also important; some farmers were able to specify exactly how many sheep and goats had been sold to finance a stretch of terrace. Within the farm, capital tends to be invested first in the most profitable parts of the enterprise (e.g. in a terrace for growing tomatoes, or a plough to expand cultivation) and to spread gradually to investments giving a lower return (e.g. the improvement of grazing land). Traders often provide start-up capital (seed, other inputs, advice) for new crops. Some 2,000 farmers in the Matuu region of Machakos began the export of Asian vegetables in the 1980s, trader interest having been stimulated by government investment in an improved road (Jaffee, 1994). Some capital is raised by groups which pool money as well as labour to generate farm improvements, non-farm businesses, or community facilities.

Again, the relative importance of different sources is not known, but it is likely that the family is the major mechanism as far as the farm is concerned. Our own study was defective in identifying the sources of capital, since it was only during the later stages that we realized its importance.

Conclusion

Overcoming blind spots concerning the importance of farm capital and the value of improved land should help experts realize that external capital merely supplements what is already being provided by farmers. However, the latter need security in land rights and access to markets to encourage investment. In Machakos, over the period 1946–85, the value of labour inputs into original terrace construction (ignoring replacement and maintenance) was an estimated 276,000 tons of maize. Government provision of tools and advice during two major programmes was equivalent to about 96,000 tons (much of which was for a type of terrace farmers later reconstructed themselves). During the 20 years in which most terraces were constructed, farmers mostly found their own tools and advice (Tiffen *et al.*, 1994: 258–9). And terracing, as we have seen, was only one of several important farm investments.

It is important that development experts recognize the value of farmers' incremental investments over time, and the diversity of their income sources, in order to identify key areas for national action. This would help avoid misplaced interventions, the rationale for which rests on a misunderstanding of the direction of environmental change. Government investments and services can promote profitable land improvements and tree-planting by farmers, especially in providing communications, a stable legal and political environment, tenure security, and information. A greater appreciation of the skill with which most farmers, male or female, juggle resources, weigh opportunities and manage and negotiate family inputs, might lead to greater acceptance of the view that they are the senior partners in rural development. Aid givers and academics should fight their inclination to nanny. There will always be the unskilful or the unfortunate, who need special help, but they should not be conceived as the majority.

11

The Cultural Construction of Environmental Policy

Paradigms & Politics in Ethiopia

ALLAN HOBEN

Introduction[1]

In the wake of the 1985 famine, the Ethiopian government launched an ambitious programme of environmental reclamation supported by donors and non-government organizations and backed by the largest food-for-work programme in Africa. Over the following five years, peasants constructed more than one million km of soil and stone bunds on agricultural land and built almost half a million km of hillside terrace. They also closed off more than 80,000 ha of hillside to most forms of use to foster the regeneration of naturally occurring plant species, and planted 300,000 ha of trees, much of it in community wood-lots.

Today, in retrospect, it is clear that much of this effort was wasted or counterproductive. The long- and short-term soil conservation benefits of the structures and trees are uncertain. The most rigorous research conducted to date shows that under most conditions terracing has lowered agricultural production instead of raising it as had been anticipated (Herweg, 1992). Farmers have been unwilling to construct or maintain structures without food-for-work or coercion, and many of the structures have fallen into disrepair. Most community wood-lots have been harvested or destroyed. Hillside closures had mixed results. Where they were built best, they tended to reduce household income from livestock, to cause environmental damage by concentrating livestock on the remaining pasture, and to harbour wild animals and pests.

[1] The research upon which this chapter is based was supported by a grant from the MacArthur Foundation to the Global Development and Environment Institute at Tufts University. Supplementary research on the Ethiopian National Conservation Strategy was carried out for the World Resources Institute. Neither organization is responsible for the views expressed here. I am grateful to Susan Hoben, James McCann and Kate Showers for comments on an earlier draft of this chapter. It is a slightly revised version of an article published in *World Development* (23 June 1995) under the title 'Paradigms and politics: the cultural construction of environmental policy in Ethiopia'.

Many factors contributed to the reclamation programme's poor performance. It was based on inadequate scientific and technical knowledge. It was implemented with a standardized approach and with little regard to regional or local agro-ecological conditions. The views and interests of the rural men and women it was intended to benefit were not solicited or heeded. Instead, implementation was top-down, authoritarian and politicized. Peasant interest in investing in long-term environmental management, to the extent that it had existed, had been undermined by the government's land reform programme.

My central concern here is not why the programme failed, but how it was thought about. My concern is with the role of a neo-Malthusian environmental policy narrative that was used by government and donors alike to justify the rapid, massive and widespread use of standardized environmental management 'packages' without research on their environmental impact or their economic costs and benefits. Understanding the context in which this happened is important in prospect as well as retrospect, for key elements of the narrative still inform thinking and planning in Ethiopia. Moreover, there is mounting evidence that the use of narratives of this type in environmental management programmes and, more generally, in many other types of development planning, is widespread and costly.

It should be emphasized from the outset that I am concerned with the common discourse about population, environment and famine in Ethiopia that was used by various actors and agencies involved in the reclamation programme. I am not primarily concerned with what individuals 'really' thought, with their personal motives, or with the details of decision-making processes. This essay is not an attempt to second-guess the experts. It is not a history of the reclamation programme, and it is not an exposé or attack on the Ethiopia and expatriate experts who designed and implemented it.

Hirschman (1968) was among the first to argue that effective development policies and programmes (i.e., ones that succeed in mobilizing funds, institutions and technology) rest on a set of more or less naive, unproven, simplifying and optimistic assumptions about the problem to be addressed and the approach to be taken. Without such a cultural script for action it is difficult for donors and aid recipients to mobilize and coordinate concerted action in the face of the many uncertainties that characterize processes of economic, political and institutional change. More recently, Roe (1991) observed that these optimistic 'enabling assumptions' are generally encoded in what he calls 'development narratives'. Although Roe does not elaborate, it is clear that the power of these narratives is enhanced through the incorporation of dominant symbols, ideologies and real or imagined historical experience of their adherents. In this sense they are culturally constructed and reflect the hegemony of Western development discourse.

To the extent that a particular development narrative becomes influential in donor community development discourse it becomes actualized in specific development programmes, projects, packages and methodologies of data collection and analysis. All of these constitute what I will gloss as the 'cultural paradigm' associated with the narrative. The cultural policy paradigm, in the sense I am using it, is thus based on concrete exemplars as well as on a set of ideas. It is not merely a set of beliefs or a theory but a blueprint for action as well.[2] Environmental rehabilitation programmes carried out in Ethiopia in the mid-1980s illustrate clearly the power and attendant problems of policy narratives and paradigms. They must be understood, however, against the background of Ethiopia's distinctive physical and institutional environment, its radical agrarian reform programme, civil strife, and the famine of 1985.

The following sections provide an account of environmental degradation, indigenous conservation, and the impact of the agrarian reform programme instituted after 1974 on peasant incentives for environmental management. Next, I analyse the genesis of the Ethiopian environmental reclamation programme in the mid-1980s. The chapter concludes by considering continuity and change in recent developments in Ethiopian environmental management.

The physical setting: fragile riches

Although Ethiopia is relatively well endowed with natural resources by African standards, its landscape shows signs of degradation interpreted by many experts as indicators of impending environmental crisis.[3] Propitious conditions for indigenous agriculture are concentrated in the highlands above 1,500 m which comprise 43 per cent of the country. Here a combination of favourable rainfall and soils has fostered the development of a variety of farming systems which has supported major concentrations of population and complex societies for several millennia.

In spite of this natural endowment, Ethiopia today is believed by experts to face an environmental crisis. Much of the North has a dissected, sloping terrain, fragile soils and is subject to highly erosive rainstorms during the main agricultural season. This area has little natural tree cover; its soils are low in organic matter and are subject to severe soil erosion. The plough-based mixed farming system contributes to soil erosion through fine tilling, mono-cropping, and a lack of vegetative cover during part of the heavy rains. On the central and southern highlands, where rainfall is higher and distributed over more of the year, soils are generally higher in organic matter. Land form is less

---

[2] The role of cultural paradigms in development planning is discussed at greater length in Hoben (1980; 1989; and 1994).
[3] For a balanced discussion of Ethiopia's natural resource base and trends in resource use, see IUCN (1990: 8–28).

rugged and there is more natural vegetative cover. In the present century natural forest, which was extensive, has been reduced and grasslands brought under the plough by a combination of conquest, spontaneous small-farmer migration, government-sponsored resettlement, expanding commercial and state farms, and private and state exploitation of the forests. Soil erosion appears to be less severe than in the North. As we shall see, the neo-Malthusian environmental narrative with which we are concerned exaggerates the rate and magnitude of degradation and misrepresents the role of human agency in causing it, but there can be no doubt that there are serious problems of soil erosion in the extensive areas of highland Ethiopia.

Indigenous environmental management before the revolution of 1974

Northern Ethiopians have long managed their landscapes and practised some forms of conservation. Historically, however, political, institutional, and economic conditions did not give landed elites or peasants strong incentives to invest in agricultural intensification and labour-intensive conservation measures, such as indigenous terracing. In any case, 'the main recurrent environmental problems in the northern highland environment were animal and human disease, crop-destroying pests, and adverse weather conditions' (McCann, 1987: 27). The relationship between society and nature in northern Ethiopia has long been characterized by flux, crisis and calamity, rather than by a harmonious balance (Mesfin, 1991). Indigenous southern Ethiopian farming systems were more sustainable before the present century, when many of them were altered by northern conquest.

The dominant secular ethos of northern Ethiopian (Amhara and Tigrean) society was military, rather than agrarian. At many times, security could not be guaranteed by the state, and warfare was not uncommon. For lord and peasant alike, the path to upward mobility lay in combat and command, not in cultivation, commerce, or the accumulation of capital. Members of the secular elite held quasi-feudal land rights (Amharic *gult*) over peasant communities that entitled them to rule and tax their subjects but not to treat the peasants as tenants or tell them how to farm. Many types of feudal land grants were of short duration. Even where grants were in principle hereditary, patterns of secession, inheritance and naming militated against the formation of transgenerational interests in improved land management. With limited exceptions there were no clearly defined ruling families grounded in the possession of particular landed estates, titles or offices.

Nobles and local notables enjoyed the taxes, tribute, labour and military services of their peasants, but their economic strategy was primarily oriented to increased extraction, not investment. Significant increases in revenue were to be secured by obtaining control over additional land grants. The path to upward mobility for the lord was thus through success in court politics, loyal service to his liege lord and military prowess, especially in the conquest of new lands. The Ethiopian landed gentry and great lords, unlike their counterparts in some other feudal societies, did not sponsor public works designed to maintain and enhance the long-term productivity of the land.

If the nobles and gentry did not have strong incentives to make long-term investments in the land, neither did the peasants. Peasants enjoyed reasonable security of access to a share of their ancestral lands, but not to particular fields.[4] The periodic reallocation of land, the scattered and changing composition of the parcels that constituted a household's holding, and the division and redistribution associated with inheritance, all militated against the concept of a family farm or homestead. There was, in northern Ethiopia, no concept of an enduring family estate or farm with its houses, lands and other assets, with an identity of its own, to be passed down through the generations. Neither was there a religious attachment to ancestral land, nor a named, status-bearing family line to be perpetuated to the glory of a father or the honour of his sons. Indeed, if they could, most men hoped to build their own homestead rather than to live in their father's house. This weak sense of familial continuity, together with the facts that land parcels might be reallocated to a distant kinsman and that farmers could not sell them for a profit or be certain of leaving them to their heirs, reduced a farmer's incentive to invest in long-term improvements in land (Hoben, 1975; Bauer, 1977).

Northern peasants have been well aware of soil erosion and traditionally have had a number of techniques for coping with it. Peasants in Wallo, for example, were familiar with bunding, terracing, contour ridging, hedging, strip cropping, ratooning and mulching (Dessalegn, 1987). They had little incentive, however, to use these techniques, especially those that required major, long-term investments, as a part of a strategy of agricultural intensification. Their control over particular fields was insecure. Agricultural production was primarily for subsistence, tax and tribute. Transport costs were high, towns few, and markets poorly developed. As one peasant remarked to the author in explaining why he did not use the more intensive techniques he had just described, 'we are lazy here because money is too expensive'.

The indigenous farming systems of southern Ethiopia did not cause serious environmental degradation. The densely populated south-western highland areas were characterized by intensive, highly integrated horticulture based on a complex mix of annual and perennial

---

[4] For a discussion of northern Ethiopian land tenure systems, see Hoben (1972), Bauer (1977), and Bruce (1976).

crops, including roots and tubers. *Ensete*, a banana-like plant with a trunk that is processed into a starchy staple food, was grown in thick stands around each homestead. Agricultural techniques included hoeing, mounding, mulching and, in some areas, terracing and irrigation. The systems were very labour-intensive. Victors in local wars sometimes claimed the labour rather than the land of those they vanquished. The south-eastern highlands of Arsi and Bali were largely populated by agro-pastoral peoples with comparatively low population densities. There were also extensive forested areas to the west and south-west in Wellega and Illubabor-Jima. Farther to the south and south-west, in the lowlands, tribal groups practised various combinations of agro-pastoralism and shifting cultivation.

The conquest and occupation of what is now southern Ethiopia by the northern Ethiopian armies in the last decades of the nineteenth century brought with it new pressures on the environment. Large numbers of northerners settled in the agro-pastoral regions, taking their plough-based agricultural system with them. In the densely settled regions a form of extractive serfdom was established. Indigenous peoples were forced to give labour and tribute to northern overlords and local notables. The introduction of the plough and of grain crops to meet northerners' demands, as well as the new labour demands, appear to have contributed to new problems of soil erosion. Northern settlers also moved southward into forested areas in the west, clearing the land and introducing plough agriculture.

On the eve of the revolution a majority of farming households in the southern highlands worked as serfs or sharecropping tenants. Tenants had little security and might owe as much as three days' labour per week or half their crops plus other gifts and services, to their landlords. In some parts of this region there were also small farmers enjoying something approaching freehold. In southern Chilalo these farmers quickly took up the improved farming practices introduced by the Swedish CADU maximum package development project (Cohen, 1987). Overall, however, the more accessible parts of the south were characterized by a very unequal distribution of land and great inequalities of status and security. In the more peripheral areas of the empire tribal groups continued to use land and pasture under more or less indigenous arrangements, except where disturbed by government or private development initiatives. The 1960s saw the beginning of large- and small-scale commercial farming in the corridors along the roads to the south and west of Addis Ababa. This process was largely stopped by the revolution and agrarian reform programme of 1974.

The impact of agrarian reform, 1974–85

Between 1974 and the late 1980s, the military regime of Mengistu Haile Mariam (the Derg) pressed an ambitious programme of agrarian reform intended to transform rural social, economic and political institutions and spur agricultural development, increase food security and address environmental problems, including deforestation and soil erosion. Ironically, the net effect of the Derg's actions was to lessen farmers' incentives for good natural-resource management by decreasing both the security of land tenure and the profitability of agriculture. At the same time it appears to have reduced, instead of increased, food security in many areas (Cohen and Isaksson, 1988).

The programme included the nationalization of natural resources, land tenure reform, the promotion of production and service cooperatives, the establishment of state farms, the imposition of production quotas, state intervention in pricing and marketing, forced villagization, large-scale long-distance resettlement, and the environmental reclamation programmes which are discussed in the next section. As the Derg struggled with a prolonged and ever more costly civil war, it also imposed taxes, required voluntary contributions and requisitioned unpaid labour, demands which often exceeded those experienced under the previous regime.

Land reform and expropriation
The Derg's land reform programme, launched in 1975, was very successful in eliminating large holdings, absentee landlordism and landlessness. It redistributed land within peasant communities on a relatively equitable basis, though it did not address inter-community or inter-regional inequalities. In most areas, land reform did not solve the problem of acute land shortage. The size of peasant holdings continued to dwindle as new households pressed their claim to land. Over time, the repeated redistribution of land and the disturbance of holdings for a series of new government programmes undermined peasants' security over particular parcels of land and decreased their incentives to use existing or new land-management practices.[5]

Under the 1975 reform, all customary and other pre-existing land rights were extinguished, and all land was declared to be public property. Individual households could farm up to 10 hectares of land. In practice they generally received much less, often less than three hectares. They had only usufruct rights over the land they cultivated, rights they could not transfer by sale, lease or mortgage. The land was subject to periodic reallocations by Peasant Associations (PAs) to balance inequalities or to accommodate new claimants. Government policies and actions also created the widespread impression that trees, in some cases including privately planted trees around homesteads, belonged to the state and could not be harvested without the permission of the authorities. The reform also abolished tenancy, agricultural wage labour, and other forms of peasant dependency on the landed classes.

[5] For a balanced account of the land reform programme, see Rahmato (1985).

Large holdings were confiscated and turned into state farms, settlement schemes, or cooperatives.

Initially, in the southern and western part of the country, land reform was welcomed by former tenants, and serfs who hoped they would have secure tenure over their holdings. In the northern highlands there was some resistance to reform by better-off peasants, though it was generally welcomed by poor and young households and landless artisans.

From the outset, however, Ethiopia's new leaders were committed to moving away from individual control over land towards producers' cooperatives (collectives) and state farms. Throughout the decade of the 1980s the central government used its increasingly top-down control to institute a series of programmes intended to move rural society in this direction and achieve other central government objectives.

The dictum that all land belonged to the state led the government to expropriate land for many of its new initiatives. Programmes that resulted in the dislocation of peasants and decreased land security included the expansion of state farms, a large-scale resettlement programme in the wake of the 1984–5 famine, the expropriation of land without compensation for government projects, and the crash agro-forestry and enclosure projects discussed in the next section.

In the last analysis, it was the arbitrary exercise of top-down authority by party cadres competing with ministry officials to institute reform that reduced the security with which peasants held their land. By the early 1980s PA leaders were regularly taking land from their members for new government programmes or requiring them to relocate their homes. There was no law or procedure concerning compensation in such cases. Over the decade the Ministry of Education evicted approximately 80,000 households for its school-building programmes. The Ministry of Coffee and Tea evicted over 15,000 households, water projects evicted 29,000, state farms over 90,000, and the Ministry of Agriculture 38,000 (for forestry and extension). These figures are dwarfed by the two million households (an estimated 8 to 10 million people) who were evicted and relocated by collectivization and villagization, and the more than one-half million households who were moved to the western lowlands in the resettlement campaign triggered by the 1984–5 drought (Dessalegn, 1991).

Production cooperatives

The most direct challenge to individual rights in land was the establishment of production cooperatives (PCs). PCs were organized with the backing of party cadres and Ministry of Agriculture officials, by 'progressive' members of a PA. They were able to appropriate the best land in each community and valuable natural resources, such as pasture land, water points and the like, for their agricultural and other enterprises. They were also able to command unpaid labour from members of their own and nearby communities. The formation of PCs thus often involved evicting large numbers of households and relocating them elsewhere, often on marginal land. Members who later wished to leave the cooperative forfeited their right to the land and other capital assets they had brought into the cooperative enterprise. PCs were unpopular among their own membership as well as among the surrounding peasantry. Though at their height cooperatives worked less than 15 per cent of the agricultural land, the programme was perceived by peasants as a major threat to their security of tenure.

All available evidence indicates that PCs were inefficient compared to individual farming in terms of productivity and resource management. Just how unpopular the enterprises were became evident when Mengistu announced a policy shift towards a mixed economy. Within a week all but a few of the nation's 3,732 registered PCs had been disbanded.

Rural economic policy

A number of economic policies decreased peasant food security. As the Derg's land policy was decreasing farmers' incentives to invest in their land, its agricultural investment and marketing and pricing policies were having the same effect by reducing the profitability of agriculture. State farms received priority in the allocation of land, machinery, credit and chemical fertilizer. Though in the late-1980s they occupied only two per cent of the cultivated land and contributed only 10 per cent of marketed agricultural production, they received approximately 70 per cent of agricultural credits. Within the peasant sector, the bulk of the Ministry of Agriculture's support went to producers' cooperatives, though their membership probably never exceeded five per cent of all peasant households (IUCN, 1990: 41). Throughout the 1980s, only 15 per cent of the government's recurrent and capital expenditures on agriculture went to peasant agriculture, including the PCs. Yet *all* studies indicate that peasant farming remained more efficient than either PCs or state farms by any measure.

From the early 1980s until the beginning of 1988, severe restrictions were imposed on private merchants. At the same time, production quotas, rigid below-market prices and an ineffective state Agricultural Marketing Corporation reduced farm-gate prices.

Falling per capita food production reduced Ethiopia's ability to cope with drought, crop disease and war. Other restrictions on peasants' coping strategies also contributed to household food insecurity. Important among these were restrictions on labour migration, share-cropping and wage labour in agriculture, restrictions on trade and attempts to discourage rural people from engaging in more than one income-generating activity. All of these measures weakened rural people's security of access to food.

Villagization

Another threat to peasants' security of land tenure was brought about by the Derg's ambitious and hurried villagization programme, which was comparable in many ways to that undertaken by Tanzania in the mid-1970s. The programme was intended to facilitate agricultural and social service delivery, social and political change, and the formation of PCs. Instead, it brought about further movement and disruption of individuals' land rights and caused many other problems, including environmental degradation, the loss of livestock through disease and reduced access to pasture, poor sanitation and the de-capitalization, especially in the South-West, of farms depending on *ensete* (false banana) and tree crops planted near the homestead.

Together all these policies appear to have depressed agricultural production and discouraged farmers from developing more intensive farming systems in the face of rising population and investing in intermediate or long-term natural-resource management (IUCN, 1990: 41–2). Indeed, it can be argued that they destabilized local food security systems and made the rural population more vulnerable than ever to drought and famine.

The environmental reclamation programme, 1985–8

In the aftermath of the drought and famine of 1985 Western donors and NGOs needed a narrative that would give them a rationale for justifying the continuation of food aid to Ethiopia, a blueprint for what to do with it, and a way to coordinate their programme with the Ethiopian regime, for which they had little regard. Under these circumstances, the idea that the underlying cause of Ethiopia's periodic famines was environmental degradation due to population increase, poverty and poor farming practices had great appeal. It provided the justification for a massive food-for-work programme supporting local-level reclamation projects. Ultimately the programme proved to be ineffective in many ways. It foundered as the Derg regime faltered and fell. Some lessons were learned from this experience, but the environmental policy narrative that informed it has proven surprisingly resilient. It is instructive to examine the narrative and the evidence that is presented to support it in some detail, as well as the nature of its appeal to the donors and the Ethiopian regime.

The narrative

The core narrative is quite simple: 'Long ago when there were fewer people in Ethiopia, indigenous farming systems and technology enabled them to make a living without seriously depleting their natural resources. Over the present century human and animal populations have grown. Indigenous farming systems have been unable to keep up. Population has exceeded carrying capacity, causing ever-increasing and perhaps irreversible environmental damage. Only a massive investment in environmental reclamation can reverse this process. People are unable to make this investment without outside assistance because they do not know how and because they are too poor to forego present for future income or to provide for their children.'

This narrative is not new, and it is not peculiar to Ethiopia. It came to play a central role in East African soil conservation and forestry policy in the 1930s (Anderson, 1984). It has been re-enunciated, reinforced, and Africanized in the wake of the environmental movement in the West, as it fits well with its interests, understandings, sentiments and with the deeply rooted Western image of Africa as a spoiled Eden.

Accompanying the basic narrative are a number of 'corollary narratives' that contextualize it in Ethiopia by 'explaining' the processes that have caused degradation and establishing the magnitude of the impending disaster. These narratives, in varying combinations, are concerned with soil, trees and water. In the past, a period seldom defined more precisely than 'before the present century', environmental degradation occurred around settlements, but communities could always move to new land, which was abundant. There was little need for conservation. The landscape was generously covered with trees, brush and grasses. A higher proportion of rain water percolated into the soil. Erosion was held in check and woodfuel was abundant, easily obtained, and cheap.

Over the present century, we are told, population growth increased, due to the partial control of epidemics and relative peace. New land was no longer easily available and fallowing periods shortened until land was under continuous cultivation. At the same time, forests are said to have been cut for firewood and agricultural expansion. A decrease in forest cover from 40 per cent to less than three per cent in the present century is an oft-repeated figure. The increasing scarcity of firewood has caused peasants to use cow-dung as a household fuel instead of using it to replace the organic matter in the fields. Steeper hillsides have been cleared and denuded of vegetation. Cultivation and overgrazing have left the soils exposed to Ethiopia's heavy rainstorms, causing severe soil erosion, reducing the nutrients available to crops and letting water run freely into the streams and rivers. Increasing livestock pressure has led to overgrazing and the deterioration of the ever-shrinking pasture land.

Though the data supporting these assertions are admittedly thin and circumstantial, powerful conclusions have been made about the rate and magnitude and direction of agro-environmental decline. A Ministry of Agriculture and FAO study in 1984 concluded that in densely settled regions of Wallo, Gondar and northern Shoa on the 'frontier' of serious degradation, soil erosion and a decline in organic matter are estimated to be reducing crop yields at a rate of two per cent per year. Based on current trends, the study describes a scenario in which by 2010 land incapable of supporting agriculture will increase from two million to 10 million hectares or 17 per cent of highland Ethiopia (Ministry of Agriculture and FAO, 1984; cited in Ståhl, 1990: 3). Another study indicates that by the same date three-quarters of all districts will be chronically food deficient (FAO/UNDP, 1984; cited in Ståhl, 1990: 3).

The actors

The neo-Malthusian environmental degradation narrative appealed to each of the major actors involved in the reclamation programme for a variety of reasons. It enabled the major Western donors, the World Food Programme (WFP), the European Economic Community (EEC), and the United States, to justify a massive food-for-work programme on the grounds that they were addressing the long-term, underlying cause of famine, rather than merely alleviating its symptoms. This enabled them to meet the criticism that they were only 'keeping people alive so that they could die in larger numbers next time the rain failed'. The donors were also able to counter the argument that the food aid would make people lazy. In addition, the narrative provided the donors with a rationale for maintaining high levels of food aid after the famine was over and for delivering food to areas that had not been affected by drought or famine.

The Reagan administration had initially opposed giving humanitarian aid on the grounds that it would strengthen a government that was violating human rights, pursuing a protracted civil war in the North, following bad economic policies, and aligned with the Soviet Union. Domestic political pressure forced them to provide food aid. Under these circumstances the reclamation programme was a least bad option, as it was narrowly technical, largely by-passed the Ethiopian government, was targeted directly on the rural poor, and would be welcomed by the growing environmental lobby in Washington.

Western NGOs were comfortable with the rationale for the reclamation programme because it fitted well with their ideals of helping people directly, teaching while helping, working with communities rather than the private sector or government, and their domestic constituency's concern with 'the environment'. Indeed the humanitarian, community and environmental emphases in the programme helped the NGOs live with, and even appreciate, the top-down and authoritarian way in which the government expedited the programme. After all, it was 'obvious' that something had to be done and that the peasants were not doing it on their own. The scale of the reclamation programme and the central role NGOs played in its implementation also meant that there were considerable organizational and financial rewards for participation.

The Ethiopian regime was hard-pressed on a number of fronts at the time of the famine. Per capita food production had been falling. The costs of the protracted civil war in the north were mounting. The regime needed food aid so that it would be able to feed the army, keep the urban population from becoming restive, and bolster its legitimacy in rural areas. The Soviet Union, on which the regime depended for arms, could not supply significant food aid or economic assistance; while the Western powers, which were in a position to help in these areas, were for ideological and strategic reasons reluctant to do so in ways that would support the war effort. Like the Western donors, the Ethiopian regime needed a common definition of the problem, a rationale for a massive food aid programme that could be conceived in narrowly technical terms and provide the basis for close cooperation.

The programme's technical components made sense to Ethiopian and expatriate experts and bureaucrats because they represented the vast expansion of a set of assumptions and an approach to soil conservation with which they were already familiar. Terracing and afforestation had been first undertaken in what was then the province of Eritrea, with the support of USAID, in the late 1960s. After the 1972–4 famine, conservation was supported by the WFP, initially involving only physical measures. In the early 1980s when the WFP first supported environmental rehabilitation programmes on a large scale, the emphasis was still on stone and earth structures.

The programme's top-down approach seemed reasonable to ministry officials and urban dwellers who traditionally have held rather negative stereotypes about peasant agriculture, intelligence and ingenuity. Farmers' reluctance to accept new practices is often attributed to their traditional attitudes, rather than to their socio-economic circumstances. For example, the fragmentation of household land holdings is attributed to inheritance rules rather than the peasant's desire to diversify his or her farming enterprise and reduce his or her risks.[6] The strengths of indigenous farming and environmental management systems are overlooked. Indeed, even when visible they are often not 'seen'.

The top-down approach also fitted well with the regime's political approach to rural development, which emphasized radical social and economic transformation, communal rather than individual incentives, and crash mobilization programmes. It was an approach that assumed peasants did not know what was good for them and would not necessarily participate in bringing about change without political agitation, education and, if necessary, coercion.

In sum, the environmental narrative; its corollary or supporting narratives; the technical package of bunds, terraces, wood-lots and closures; and the top-down, crash mobilization programme approach to implementation made sense to all of the key actors except, perhaps, to the peasants, who were not asked. They, too, proved willing to go along with the programme, and, at times, to agree with the narrative because they appreciated the grains and edible oils they received at well over

---

[6] For a discussion of reasons for fragmentation, see Bruce *et al.* (1994: 28–29).

market wage rates and because they had learned that it was politic to agree with official views.

Problems with the narrative

There is little doubt that the population in Ethiopia is increasing at over 2.5 per cent per year, that many Ethiopians are poor, hungry and vulnerable to famine, and that soil erosion is a serious problem in many highland areas. What is in question is whether the neo-Malthusian environmental degradation paradigm provides an adequate framework for understanding the genesis of these phenomena, their causal interrelations, or what should be done to alleviate them. To anticipate, I will argue that there are major difficulties with the narrative and its supporting 'data stories'. They misrepresent what has happened, often exaggerating the rate of degradation and the ways in which human activity is causing it. They preclude the examination of alternative hypotheses based on the experience of other countries and inhibit scientific inquiry. They have contributed to a massive investment of funds, time and effort in activities that have at best had a marginal beneficial impact. These problems with the paradigm for action are addressed in the next section, while this section addresses problems with the core narrative and its supporting data.

Difficulties with the Ethiopian data
Famine is not a new phenomenon in Ethiopia. It was first recorded in the ninth century, and ten major famines occurred within the two centuries following the expulsion of Muslim invaders in 1540. The great famine of 1888–9 caused widespread death and devastation. More localized but not less lethal famines occurred in 1916–20, 1927–8, and 1934–5. This recurring pattern of famine in itself calls into question the narrative that attributes it primarily to recent environmental decline.

There are also a number of difficulties with the thesis that the deforestation of the highland plateau has played a major role in recent environmental degradation and famine. A careful review of historical sources reveals no significant change in tree cover in the northern highland landscape over the past century and a half, though there has been a decline in natural forest in some of the canyons that dissect the plateau.[7] The loss of natural forest, which is serious though overstated, has occurred in the southern half of the country, while the denuded hillsides and barren landscapes in the narrative are found only in the northern regions, where the FAO report locates the 'frontier of serious degradation'.

Another difficulty with the loss of tree cover thesis is that on much of

[7] For an excellent discussion of agricultural history in northern Ethiopia, see McCann (1995), Chapter 4.

the northern plateau eucalyptus trees have been integrated into the farming system since they were introduced in the last decades of the nineteenth century. In much of the north it is hard to find a homestead that lacks a stand of these trees and does not use them for construction and a part of its fuel needs. In the past two decades an imported species of juniper and bamboo have also been extensively planted around house sites in Dega Damot, Gojjam where I conducted research in the early 1960s. Indeed, when I went back to the district after almost two decades, what struck me most vividly was the increase in tree cover on the farmed high plateau. The increase of tree cover with population pressure is not surprising and has been observed in Kenya (Shipton, 1989).

The fact that trees have been integrated into highland farming systems spontaneously without government extension programmes calls into question the narrative that says peasants lack the ability or foresight to plant trees without environmental education, training, and access to subsidized seedlings from nurseries. Ironically, the surge in concern with fuelwood as a national issue can be traced back to the mid-1970s, when peri-urban plantations of eucalyptus declined sharply shortly after they were nationalized by the Derg. A great increase in tree planting on individually controlled land was noted by NGO workers in central and southern Ethiopia in 1990–1, immediately after the Derg abandoned key features of the agrarian reform programme and relaxed control over the private sector.

There are several problems with the dung-burning thesis. Burning dung is not a new practice. McCann's historical study of agriculture in Ethiopia finds that it was done in the same areas today as in the nineteenth century. This casts doubt on the fuelwood-shortage theory of increased dung use. Dung is a preferred fuel for certain types of cooking. It is often used for fuel even by better-off farmers who have access to fuelwood, because of its combustive properties and because it is cheaper (Pankhurst, 1992: 90–6). The nitrogen loss due to the use of dung as fuel instead of spreading it on farmland may well also be over-estimated, as the nitrogen in dung is very volatile. Nor does burning dung as a domestic fuel lead to its complete loss as a fertilizer, for burning lowers the pH and concentrates any phosphorus present. Ashes from the homestead are invariably spread on the garden and adjacent fields. Indeed, in many of the areas that are heavily dependent on dung for fuel, farmers also pile up and burn sod prior to planting their crops, apparently for the same reasons.

Widely accepted estimates of the reduction in crop yields caused by erosion and biological degradation (a decline in organic matter), such as those of the FAO/UNDP (1984) report, are almost certainly much too high. A recent re-analysis of the data by Peter Sutcliffe, former Senior Technical Advisor to the National Conservation Secretariat in Ethiopia, indicates that loss of crop yield estimates are from 10 to 15 times too high (Sutcliffe, personal communication). Over-estimation of losses in production due to soil erosion are commonplace in Africa, due in part to methodological problems (Stocking, this volume). In the last analysis, the data on soil erosion and nutrient loss in Ethiopia is quite thin.

Questions not asked
The influence of the narrative has largely prevented planners from examining counter-narratives or alternative explanations of Ethiopia's environmental and food-security problems and from adequately testing reclamation technologies and approaches.

The neo-Malthusian narrative's denigration of indigenous agriculture has led experts and planners to overlook and filter out much information about the strengths of indigenous resource-management practices. There is virtually no mention of the fact that agro-forestry is almost universal in highland farming systems. There is little discussion of indigenous techniques of soil amelioration, including manuring, spreading ashes from manure which has been burned, and the use of leguminous crops in rotation, except for the occasional claim that these practices are dying out. There is little discussion or even acknowledgement of indigenous terracing, which is extensive in some areas, or of indigenous run-off ponds, or irrigation. Densely settled areas in the south-west are always said to be at environmental risk, with no investigation of the distinctive farming systems that appear, in some cases, to have sustained such densities for centuries.

Equally damaging, the neo-Malthusian narrative rests on an essentially undynamic view of peasant behaviour, one in which it is only possible to make linear projections of the rate at which Ethiopia is heading towards environmental collapse. What planners need to know is under what circumstances peasants do and do not use the resource-management techniques of which they are aware, so that policies and investments can help to make these circumstances more common. What, for example, were the effects of the traditional political and economic organization of the empire? What were the effects of the changes in the political economy brought about by Imperial regime in the twentieth century?[8] What were the effects of the Derg's programme of agrarian reform, limitations on the private grain trade and restrictions on population moment? What were the effects of years of mobilization for war? How did all of these changes affect particular groups of rural Ethiopians' incentives to manage the environment and the access to food? Asking these kinds of questions would focus attention beyond the physical processes taking place on the land and broaden our inquiry to include questions about transportation, marketing and trade, about food processing and storage, about land tenure and land markets, and about

[8] For an excellent analysis of the linkages between changing government policy, environmental management, food and population movement in the first half of the present century, see McCann (1987).

the relationship of many of these factors to demographic issues and family planning. Indeed, one could develop a counter-narrative that would 'explain' Ethiopia's environmental crisis in terms of a failure to intensify agriculture, resulting from political insecurity and bad policy, or alternatively from too little infrastructure and capitalism.

Problems with the paradigm for action and its implementation

From a practical standpoint, the worst thing about the neo-Malthusian narrative is that it fostered a major investment in technologies and activities that did little to address environmental degradation or farmers' needs. In the wake of the famine, existing government reclamation programmes were greatly expanded. Activities on peasant lands were organized by the Community Forestry and Soil Conservation Development Department of the Ministry of Agriculture, which was responsible for all food-for-work programmes. The effort was supported by the WFP's Project 2488, the EEC, and United States donations of grain and edible oils. Other donors provided technical equipment and tools. Non-governmental organizations played a major role in implementation, each being assigned to a particular geographical location.

Project sites were selected by extension workers from the Ministry of Agriculture. The peasants were organized in large work teams to do the actual digging, pitting and planting. Generally this was done by contracting with an individual who was responsible for hiring the workers and making sure the work was done on a particular section of the project. The peasants contracted for the job were entitled to a daily payment of 2–3 kg in wheat and 120 g of edible oil to be delivered each month on the basis of the project work completed.

The programme grew rapidly to become the second largest food-for-work programme in the world and the largest in Africa. At its height the programme was active in nine regions and provided 100,000 tonnes of food to up to 800,000 people. By the beginning of 1990 the peasants had constructed more than one million km of bunds on farm land and had terraced almost half a million km of hillside. In addition, 80,000 ha of hillside had been enclosed and it was claimed that 300,000 ha had been planted to trees (Ståhl, 1990: 5). Hundreds of tree nurseries had been established with the capacity to produce an astounding 100 million seedlings per year.

The decision by all parties to proceed with the reclamation programme was based on certainty that the explanatory model of degradation was correct, that the peasants were unable and unwilling to take action without outside support, and that they had no indigenous knowledge or techniques for managing their environment that were worth taking into account. Based on these assumptions, donor willingness to implement the programme and let the government spearhead the effort, using its top-down and authoritarian approach to rural development and administration, seemed to make sense. After all, 'the problem was urgent'. It was obvious what had to be done and done quickly. There was no time to wait for time-consuming research. In retrospect, all these assumptions are questionable.

Between 1985 and 1990 the programme encountered a number of difficulties and was increasingly criticized by members of the NGO community and experts in the Ministry of Agriculture. Many critics complained about the way in which the programme had been implemented. Some found fault with various components of the package. Few questioned the validity or adequacy of the underlying environmental narrative.

In their eagerness to expand their programmes and dispense food stocks, donors and NGOs were not always able to ensure that their activities focused on localities that were experiencing unusual food deficits or environmental problems. Indeed, in reading project design documents, one is struck by the extent to which the environmental degradation narrative was mapped onto the local landscape to justify uniformly the need for food and conservation work.

By the end of the 1980s, experience and evaluations had revealed a number of fundamental technical and organizational problems with the rehabilitation activities. An impact evaluation carried out for the WFP and the Ministry of Agriculture reported mixed views among peasants (Yeraswork, 1988). Many said they liked the soil bunds which retained soil and moisture. Most complained that the stone bunds and terraces reduced arable land and harboured rodents. Some peasants complained that terracing reduced yields by raising subsoil to the surface, making it hard to plough, and reducing field size. Some peasants complained bitterly that the terraces and *fanyaju*, a special type of terrace designed to channel rainwater run-off, increased problems of soil erosion. Their complaints were generally ignored by local authorities.

Meanwhile, a major project of research and experimentation in soil conservation measures at seven experimental research stations in different agro-ecological zones was begun in 1981.[9] In time it was found that farmers did not find the physical conservation works attractive, and additional social and economic research was undertaken. By 1991 it had become clear that production on control plots was significantly higher under most crops and conditions than it was under any of the conservation measures (Herweg, 1992). In other words, contrary to expert opinion and to what the extension agents had been telling

---

[9] The Soil Conservation Research Project was carried out with Swiss funding and technical support from the University of Berne. Its methodology is described in Herweg and Grunder (1991).

farmers, the conservation measures lowered production, income and food security. The main contributing factors found by the study were: that 10–20 per cent of cropped area, even more on steep slopes, was lost to the conservation structures; the infestation of bunds by rodents whose habitat would normally be destroyed by ploughing; weeds; water-logging; and the difficulty of ploughing and threshing in narrow spaces (Herweg, nd: 11–12). These were, of course, the same problems of which peasants had complained in the WFP/Ministry of Agriculture study (Yeraswork, 1988).

Hillside closures generally achieved impressive results in terms of vegetative regeneration, but species unpalatable to livestock and trees tended to dominate fodder grasses. This intensified destructive grazing on hillsides adjacent to enclosures (Fruhling, 1988; Hultin, 1988).

Community forestry was not popular or, in the long run, successful. In areas of land hunger farmers resented setting aside land for community forestry. In areas of the south where naturally occurring trees were still abundant, farmers could see little point in planting trees. Where wood-lots were planted farmers often complained that the trees reduced agricultural production through shading, root interference and attracting anti-crop wildlife, such as birds (CRDA, 1990a: 6). It was also unclear who would benefit from the wood-lots, which were viewed by the peasants as belonging to the NGOs or the state. Nor did their experience lead them to believe otherwise, as they had to get permission from ministry officials to harvest their trees. It is not surprising, then, that tree-lots were not tended or guarded well, that sapling survival rates were low, and that farmers generally refused to work on community forestry or other rehabilitation activities without continuing payments in food.

The magnitude of farmer displeasure with community wood-lots and many other communal aspects of government policy was brought out clearly in the events that followed a speech by Mengistu Haile Mariam in March 1990. The speech was widely interpreted by farmers to mean that they were free to dissolve production cooperatives, repossess their former lands, live where they pleased and generally ignore the more onerous aspects of the agrarian reform programme. In the weeks and months that followed there were widespread reports of farmers cutting trees and uprooting seedlings for a variety of reasons, including the desire to reclaim land lost to production cooperatives, community wood-lots and major reforestation projects; the desire to expand land under production or to reduce the negative effect of trees on adjacent fields; and the belief that by clearing and cultivating land they might establish a claim to long-term use or ownership.[10]

In the last analysis, only a small part of the degraded highlands near

[10] The information was gathered through field trips and interviews with researchers from Addis Ababa University, representatives of the Ministry of Agriculture, NGOs, and major donor organizations active in environmental rehabilitation programmes.

the roads had been 'rehabilitated', and even in these areas, activities which farmers considered useful, such as soil bunding, did not appear to be sustainable. Farmers appreciated the food but were not willing to sustain the effort on their own initiative.

After 1990, as the Derg began to loosen its grip and crumble, critics began to complain openly about the top-down approach to the programme. Participants at an NGO-sponsored workshop on community forestry held in October 1990 drew attention to the programme's reliance on government-imposed institutions and their leaders. They pointed out that small farmers generally were not involved in identifying their own needs and problems, establishing priorities, evaluating alternative solutions, or planning how they were to be implemented. As a result, indigenous farming systems, technical knowledge, and common property institutions were ignored, and farmer incentives for participating in community forestry projects were poor (CRDA 1990a; 1990b).

The post-Derg reform period, 1991 to the present

In May 1991 rebels from Tigray toppled the Derg, occupied Addis Ababa and formed a new Transitional Government of Ethiopia (TGE). The new government espoused a policy of ethnic self-determination and decentralization. It has subsequently introduced formal changes in administration and governance to implement these policies, but these have been countered, to a large extent, by the re-emergence of a one-party state under the stewardship of the Tigrean-dominated Ethiopian People's Revolutionary Democratic Front (EPRDF).

The new regime remains committed to addressing environmental problems. While it is considered pro-peasant and stresses the importance of local people's 'participation' in all its programmes, it is not clear to what extent it has escaped old orthodoxies. In its home area, Tigray, where it enjoys great popular support, it has actively promoted terracing and reforestation projects by moblizing local peasant associations. In the rest of the country, where governance has been weak, the reclamation programme has stagnated. The regime has not been eager to pursue unpopular programmes, not even the collection of rural taxes.

The regime also maintained the previous government's commitment to preparing a National Conservation Strategy (NCS) with the assistance of the International Union for the Conservation of Nature (IUCN, 1990). The changes they introduced into the process by which this was done and the way the process eventually played itself out are instructive. For despite a conscious effort to obtain peasant knowledge, the environmental policy narrative still exerts a strong influence over thinking about environmental management.

Work for the first phase of the NCS was carried out through two missions in 1989 and the Phase I report was completed in March 1990. When the second phase of NCS planning was initiated later that year, it was envisaged that, following a conference attended by experts and officials from relevant ministries, the secretariat would spend approximately 12 months drafting the strategy. It was only late in 1991, after the establishment of the TGE, that it was decided to take a bottom-up and decentralized regional approach. This approach fits with the new regime's policy and with the expatriate advisor's belief that such a bottom-up and participatory procedure would enable the NCS to tap peasants' ideas and practices concerning environmental problems and management.

It was anticipated that in each region work would be undertaken in four stages: stock-taking and resource assessment; identification of key resource issues, problems and potentials and framework for development; a regional conference to aggregate sub-regional findings and develop a regional strategy; and the preparation of an action plan and an investment plan. It was only after these stages had been completed in the regions that the national plan was to be drawn up, using the regional plans, studies conducted by members of the secretariat, and any additional studies available in the process. As things turned out, there were many logistical difficulties in conducting the regional and local assessments. Peasant input into the process was nominal at best. In the end, the secretariat was pressured to complete the NCS for an international seminar before any of the regions had completed their strategies.

A careful reading of the voluminous zonal assessments that were produced as a part of the NCS exercise is instructive. It indicates that negative views of peasant agriculture held by local task force members filtered out much information about the strengths of indigenous resource management practices, even though this was explicitly called for in the guidelines developed by the secretariat. There is, for example, virtually no mention of indigenous agro-forestry. There is little discussion of indigenous techniques of soil amelioration. There is little discussion or even acknowledgement of indigenous terracing or of indigenous run-off ponds, or irrigation. Densely settled areas in the south-west are said to be at environmental risk because of population pressure with no investigation of the distinctive farming systems that appear until recently to have sustained such densities for centuries. Similar gaps in regard to the strength of indigenous practice are embedded in some of the reports done by experts for the Ethiopian Forestry Action Plan (EFAP), an ambitious and in many ways excellent planning effort that was incorporated into the NCS. Perhaps the most striking example of this in the EFAP is the estimation of a woodfuel deficit based on estimates that do not include the major source of peasants' woodfuel: on-farm agro-forestry. In sum, ideas and 'facts' from the old narrative were used uncritically, while new information and alternative understandings were screened out.

Conclusion

In this chapter I have illustrated the way a particular neo-Malthusian narrative came to play a central role in the way donors, the Ethiopian government, and NGOs conceptualized the causes of the 1985 famine and attempted to address its underlying causes with food aid. I have argued that the narrative and its corollaries enabled these powerful organizations with divergent interests and values to overcome their differences, craft a common policy and coordinate their activities. The resulting crash programme's success in mobilizing and deploying resources for environmental reclamation in Ethiopia was not matched by success in meeting rural people's agricultural needs, or halting degradation. In retrospect it seems clear that addressing Ethiopia's environmental problems will require better scientific and technical research, a region- and site-specific approach, and the active involvement of rural stakeholders in all stages of programme development and implementation.

I have also argued that the neo-Malthusian narrative was not new and has not been seriously challenged. It was current in Ethiopia before the famine and fitted well with traditional Ethiopian elite attitudes towards the peasantry. It has survived the fall of the Derg and the establishment of a new regime. It is likely to shape the way Ethiopian environmental problems are addressed in the future.

Ethiopia and the 1985 famine are in many ways unique. What lessons, then, can we draw from this case study? First, it seems clear, from this and the other essays in this volume, that the underlying problems with the use of narratives in paradigms in planning for environment and development are widespread and enduring. The rise and decline of the woodfuel crisis, the shifting sands of desertification policy, and the history of thinking about pastoral livestock programmes all illustrate the way narratives have been mapped onto diverse African landscapes.

It is evident, however, that the use of narratives in planning for environment and development causes more difficulties under some conditions than others.

What can be said of these conditions? The Ethiopian case suggests that the power of a development narrative varies with the context in which it is used. Generally, it is enhanced when: (1) donor experts and their domestic constituencies are strongly attached to them; (2) there is political, strategic or moral pressure on donors to act quickly; (3) there has been little technical and socio-economic research, and there is a weak data base on the problem in the recipient nation; (4) the recipient country must rely heavily on expatriate experts for advice; (5) the recipient government is dependent on foreign assistance; and (6) the recipient government, being weak or authoritarian or both, does not have the institutional capacity to hear and learn from its rural people.

If these generalizations are correct there may be reason for cautious optimism. To the extent that development assistance is less abundant, African governments are more democratic in the broadest sense of the term, and African scientific capacity is enhanced, policy narratives may become more realistic and may be challenged more quickly when they are not.

Abel, N.O.J., 1993, 'Reducing cattle numbers on Southern African Communal Range: Is it worth it?', in R.H. Behnke, I. Scoones and C. Kerven (eds), *Range Ecology at Disequilibrium*. Overseas Development Institute, London, pp. 173–95.

Abel, N.O.J. and Blaikie, P.M., 1989, 'Land degradation, stocking rates and conservation policies in the communal rangelands of Botswana and Zimbabwe', *ODI Pastoral Development Network Papers* 29 (a). Overseas Development Institute, London.

Abel, N. and Stocking, M.A., 1987, 'A rapid method for assessing rates of soil erosion and sediment yields from rangelands', *Journal of Range Management* 40: 460–6.

Acocks, J.P.H., 1975, *Veld Types of South Africa. Memoirs of the Botanical Survey of South Africa* 40, (Pretoria, second ed.). Botanical Research Institute, Department of Agricultural Technical Services, Republic of South Africa.

Adam, J.G., 1948, 'Les reliques boisées et les essences des savanes dans la zone préforestière en Guinée française', *Bulletin de la Société Botanique Française* 98: 22–6.

Adam, J.G., 1968, 'Flore et végétation de la lisière de la forêt dense en Guinée', *Bulletin d'IFAN* Série A 30 (3): 920–52.

Adams, W.M., 1988, 'Rural protest, land policy and the planning process on the Bakolori Project, Nigeria', *Africa* 58 (3): 315 36.

Adams, W.M., 1992, *Wasting the Rain: Rivers, People and Planning in Africa.* Earthscan, London.

Adams, W.M. (in press), 'Capture and désengagement: indigenous irrigation and development in sub-Saharan Africa', in L.J. Slikkerveer and W.M. Adams (eds), *Indigenous Irrigation and Change in African Agriculture*, University of Leiden.

Adams, W.M. and Kimmage, K., 1992, 'Wetland agricultural production and river basin development in the Hadejia-Jama'are valley, Nigeria', *Geographical Journal* 158: 1–12.

Adams, W.M., Potkanski, T.P. and Sutton, J.E.G., 1994, 'Indigenous farmer-managed irrigation in Sonjo, Tanzania', *Geographical Journal* 160: 17–32.

Adams, W.M., Watson, E.E. and Mutiso, S.K. (in prep. a), 'Rules, theft and gender: water rights in an indigenous irrigaton system, Marakwet, Kenya' (ms submitted to *Development and Change*).

Adams, W.M., Watson, E.E. and Mutiso, S.K. (in prep. b), *Running Water: Indigenous Irrigation and Development in Marakwet, Kenya.*

Adelodun, M., 1944, 'Firewood supply from Kano farms', *Farm and Forest* 5 (2): 30.

African Development and Economic Consultants Ltd (ADEC), 1986, 'Machakos Integrated Development Programme Socio-Economic Survey: Final Report. Volume 1: main report', Mimeo from ADEC, Nairobi to MIDP, Ministry of Planning and National Development, Machakos.

Alexander, J., 1991, 'The unsettled land: the politics of land redistribution in Matabeleland, 1980–1990', *Journal of Southern African Studies* 17: 581–610.

Allan, W., 1965, *The African Husbandman*. Oliver and Boyd, Edinburgh.

Amanor, K.S., 1994a, *The New Frontier: Farmer Responses to Land Degradation*. UNRISD, Geneva and Zed Books, London.

Amanor, K.S., 1994b, 'Ecological knowledge and the regional economy: environmental management in the Asesewa District of Ghana', *Development and Change* 25 (1): 41–67.

Anderson, D., 1987, *The Economics of Afforestation: A Case Study in Africa*. Johns Hopkins University Press, Baltimore, MD.

Anderson, D.M., 1984, 'Depression, dust bowl, demography, and drought: the colonial state and soil conservation in East Africa during the 1930s', *African Affairs* 83 (332): 321–43.

Anderson, D.M., 1988, 'Cultivating pastoralists: ecology and economy among the Il Chamus of Baringo 1840–1980', in D.H. Johnson and D.M. Anderson (eds), *The Ecology of Survival: Case Studies from Northeast African History*. Lester Cook Academic Publishing, London.

Anderson, D.M., 1989, 'Agriculture and irrigation technology at Lake Baringo', *Azania* 24: 89–97.

Anderson, D.M. and Grove, R., 1987, 'The scramble for Eden: past, present and future in African conservation', in D.M. Anderson and R. Grove (eds), *Conservation in Africa: People, Policies and Practice*. Cambridge University Press, Cambridge, pp. 1–12.

Anderson, D.M. and Millington, A.C., 1987, 'Political ecology of soil conservation in anglophone Africa', in A.C. Millington, S.K. Mutiso and J.A. Binns (eds), *African Resources Vol. 2: Management*. Department of Geography, University of Reading, pp. 48–59.

Anglo-French Commission, 1973, 'Rapport de la Mission Forestière Anglo-Française Nigeria-Niger (Decembre 1936–Fevrier 1937)', *Bois et Forêts des Tropiques* 148: 3–26.

Anon, 1925, 'The dangers of soil erosion and methods of prevention', *Rhodesian Agricultural Journal* 22: 533–42.

Anon, 1931a, 'Editorial. Grass pastures', *Rhodesian Agriculture Journal* 28: 126–7.

Anon, 1931b, 'Pasture research in Southern Rhodesia: Farmers' Day at Marandellas', *Rhodesian Agriculture Journal* 28: 619–23.

Anon, 1990, 'EEC/FGN Katsina Afforestation Project: brief information about the project', *The Greenlight Magazine* 2 (2): 11–20.

Anon, n.d., 'Outreach Programme for the Mkomazi Project, Manifesto' (authorship not stated). Archive for Mkomazi Research Programme, 1994–1996, Royal Geographical Society, London.

Ansty, D., 1955, 'Mkomazi game reserve', *Oryx* 3: 183–5.

Areola, O., 1987, 'The political reality of conservation in Nigeria', in D.M. Anderson, and R. Grove (eds), *Conservation in Africa: People, Policies and Practice*, Cambridge University Press, Cambridge, pp. 277–92.

Aubréville, A., 1949, *Climats, forêts et désertification de l'Afrique tropicale*. Société d'Edition de Géographie Maritime et Coloniale, Paris.

Avenard, J.-M., Bonvallot, J., Latham, M., Renard-Dugerdil, M. and Richard, J., 1974, *Aspects du contact forêt-savane dans le centre et l'ouest de la Côte d'Ivoire: étude descriptive*. ORSTOM, Abidjan.

Bah, M.O., 1989, 'Eaux et forêt, règlement de la chasse, protection de la faune', Interview, *Guinée Agricole* 5: 23–5, MARA, Conakry.

Bah, M.O., 1990, *Construire la Guinée après Sékou Touré*. L'Harmattan, Paris.

Bassett, T.J., 1993, 'Introduction: the land question and agricultural transformation in Sub-Saharan Africa', in T.J. Bassett and D.E. Crummey (eds), *Land in African Agrarian Systems*, University of Wisconsin Press, Madison, WI, pp. 3–34.

Bassett, T.J., 1993b, 'Land use conflicts in pastoral development in Northern Côte d'Ivoire', in T.J. Bassett and D.E. Crummey (eds), pp. 131–54.

Bauer, D.F., 1977, *Household and Society in Ethiopia*. African Studies Center, Michigan State University, East Lansing.

Bdliya, H., 1991, 'Complementary land evaluation for small-scale farming in Northern Nigeria', *Journal of Environmental Management* 33 (2): 105–16.

Behnke, R.H., 1985, 'Measuring the benefits of subsistence versus commercial production in Africa', *Agricultural Systems* 16: 109–35.

Behnke, R.H. and Scoones, I., 1993, 'Rethinking range ecology: implications for rangeland management in East Africa', in R.H. Behnke, I. Scoones and C. Kerven, (eds), *Range Ecology at Disequilibrium,* Overseas Development Institute, London, pp. 1–30.

Behnke, R.H., Scoones, I. and Kerven, C. (eds), 1993, *Range Ecology at Disequilibrium: New Models of Natural Variability and Pastoral Adaptation in African Savannas*. Overseas Development Institute, London.

Beinart, W., 1984, 'Soil erosion, conservationism and ideas about development: a southern African exploration, 1900–1960', *Journal of Southern African Studies* 11 (1): 52–83.

Beinart, W., 1993, 'The night of the jackal: sheep, pastures and predators in South Africa, 1900–1930', unpublished paper, preliminary version published in *Revue Française d'Histoire d'Outre-mer*, LXXX (298): 105–29.

Beinart, W., 1994, 'Farmers' strategies and land reform in the Orange Free State', *Review of African Political Economy* 21 (61): 389–402.

Bell, R.H.V., 1987, 'Conservation with a human face: conflict and reconciliation in African land use planning', in D.M. Anderson and R. Grove (eds), *Conservation in Africa: People, Policies and Practice*. Cambridge University Press, Cambridge, pp. 79–101.

Berger, D.O., 1993, *Wildlife Extension: Participatory Conservation by the Maasai of Kenya*. ACTS Press, Nairobi.

Bergmann, H. and Boussard, J.M., 1976, *Guide to Economic Evaluation of Irrigation Projects (revised version)*. Organization for Economic Cooperation and Development (OECD), Paris.

Berkes, F. (ed.), 1989, *Common Property Resources: Ecology and Community-Based Sustainable Development*. Belhaven Press, London.

Berry, L., 1983, *Assessment of Progress in the Implementation of the Plan of Action to Combat Desertification in the Sudano-Sahelian Region 1977–1984*. UN Sudano-Sahelian Office, New York.

Biot, Y., 1990, 'The use of tree mounds as benchmarks of previous land surfaces in a semi-arid tree savanna, Botswana', in J.B. Thornes (ed.), *Vegetation and Erosion*, Wiley, Chichester, pp. 437–50.

Biot, Y., 1993, 'How long can high stocking densities be sustained?', in R.H. Behnke, I. Scoones and C. Kerven (eds), *Range Ecology at Disequilibrium*. Overseas Development Institute, London, pp. 153–72.

Biot, Y., Lambert, R. and Perkin, S., 1992, 'What's the problem? An essay on land degradation, science and development in sub-Saharan Africa', *Discussion Paper* 22. School of Development Studies, University of East Anglia.

Bird, A., 1983, 'The land issue in large scale irrigation projects: some problems from Northern Nigeria', in A.T. Grove and W.M. Adams (eds), *Irrigation in Africa: Problems and Problem Solving*, African Studies Centre, Cambridge University.

Boserup, E., 1965, *The Conditions of Agricultural Growth: The Economics of*

Agrarian Change Under Population Pressure. Allen and Unwin, London (Reprinted by Earthscan, 1993).

Boserup, E., 1970, 'Present and potential food production in developing countries', in W. Zelinsky, L.A. Kosinski and R.M. Prothero, *Geography and a Crowding World.* Oxford University Press, Oxford.

Botkin, D.B., 1990, *Discordant Harmonies: A New Ecology for the Twenty-first Century.* Oxford University Press, New York.

Boutrais, J., 1992, 'L'élevage en Afrique tropicale: une activité dégradante?' *Afrique contemporaine* 161: 109–25.

Brokensha, D., Warren, D.M. and Werner, O. (eds), 1980, *Indigenous Knowledge Systems and Development.* University Press of America, Lanham, MD.

Bromley, D.W. (ed.), 1992, *Making the Commons Work: Theory, Practice and Policy.* Institute for Contemporary Studies Press, San Francisco, CA.

Bromley, D.W. and Cernea, M.M., 1989, 'The management of common property resources: some conceptual and operational fallacies', *World Bank Discussion Papers* 57. World Bank, Washington, DC.

Bruce, J.W., 1976, 'Land reform and indigenous communal tenures: The case of *chiguraf-gwoses* in Tigray, Ethiopia', dissertation, University of Wisconsin-Madison, Madison, WI.

Bruce, J.W., Hoben, A. and Rahmato, D., 1994, 'After the Derg: an assessment of rural land tenure issues in Ethiopia', mimeo, Land Tenure Center, University of Wisconsin-Madison, Madison, WI.

Buchanan, K.M. and Pugh, J.C., 1955, *Land and People in Nigeria.* University Press, London.

Burgess, J., 1990, 'The production and consumption of environmental meanings in the mass media: a research agenda for the 1990s', *Transactions of the Institute of British Geographers* N.S. 15: 139–61.

Campbell, B., DuToit, R. and Attwell, C., 1989, 'The Save study: relationships between the environment and basic needs satisfaction in the Save catchment, Zimbabwe', Supplement to *Zambezia*, University of Zimbabwe, Harare.

Carney, J., 1992, 'Contract rice farming and female rice growers in the Gambia', *ODI Irrigation Management Network Africa Edition* Paper 15, July. Overseas Development Institute, London.

Casta, P., Chopart, J.-L., Janeau, J.-L. and Valentin, C., 1989, 'Mesure du ruissellement sur un sol gravillonnaire de Côte d'Ivoire après six ans de culture continue avec ou sans labour', *L'Agronomie Tropicale* 44 (4): 255–62.

Caughley, G., Shepherd, N. and Short, J. (eds), 1987, *Kangaroos: Their Ecology and Management in the Sheep Rangelands of Australia.* Cambridge University Press, Cambridge.

Cernea, M.M. and Meinzen-Dick, R., 1994, 'Design for water users associations: organisational characteristics', *ODI Irrigation Management Network* Paper 30. Overseas Development Institute, London.

Chambers, R., 1983, *Rural Development: Putting the Last First.* Longman, London.

Chambers, R., 1990, 'Microenvironments unobserved', *Gatekeeper Series* SA22. International Institute for Environment and Development (IIED), London.

Chambers, R., 1993, *Challenging the Professions: Frontiers for Rural Development.* Intermediate Technology Publications, London.

Chambers, R., Pacey, A. and Thrupp, L.A. (eds), 1989, *Farmer First: Farmer Innovation and Agricultural Research.* Intermediate Technology Publications, London.

Chevalier, A., 1928, 'Sur la dégradation des sols tropicaux causée par les feux de brousse et sur les formations végétales régressives qui en sont la conséquence', *Comptes Rendus de l'Académie de Sciences* CLXXXVIII: 84–6.

Christian Relief and Development Association (CRDA), 1990a, Proceedings of the Community Forestry Development Workshop, Addis Ababa, 4–5 April.

Christian Relief and Development Association (CRDA), 1990b, Proceedings of the Workshop on Environment Impact Assessment, Addis Ababa, 15–16 October.

Clay, E., and Schaffer, B. (eds), 1984, *Room for Manoeuvre: An Exploration of Public Policy in Agriculture and Rural Development*. Heinemann, London.

Cleaver, K., 1992, 'Deforestation in the western and central African forest: the agricultural and demographic causes, and some solutions', in Cleaver, K. *et al.* (eds), *Conservation of West and Central African Rainforests*, World Bank Environment Paper No.1, World Bank, Washington, DC, pp. 65–78.

Clements, F.E., 1916, 'Plant succession: an analysis of the development of vegetation', *Carnegie Institute Publications* 242: 1–512.

Cline-Cole, R., 1994, 'Political economy, fuelwood relations, and vegetation conservation. Kasar Kano, Northern Nigeria, 1850–1915', *Forest and Conservation History* 38: 67–78.

Cline-Cole, R., 1995, 'Livelihood, sustainable development and indigenous forestry in dryland Nigeria', in T. Binns (ed.), *People and Environment in Africa*, Wiley, Chichester, pp. 171–85.

Cline-Cole, R., Main, H.A.C. and Nichol, J.E., 1990, 'On fuelwood consumption, population dynamics and deforestation in Africa', *World Development* 18 (4): 513–27.

Cohen, J., 1987, *Integrated Rural Development: The Ethiopian Experience and the Debate*, Scandinavian Institute of African Studies, Uppsala.

Cohen, J.M. and Isaksson, N.-I., 1988, 'Food production strategy debates in revolutionary Ethiopia', *World Development* 16 (3): 323–48.

Collet, D., 1987, 'Pastoralists and wildlife: image and reality in Kenya Maasailand', in D.M. Anderson and R. Grove (eds), *Conservation in Africa*, Cambridge University Press, Cambridge, pp. 129–48.

Collingridge, D. and Reeve, C., 1986, *Science Speaks to Power: The Role of Experts in Policy Making*. Frances Pinter, London.

Conable, B., 1987, Address as prepared for delivery to the World Resources Institute, Washington, DC, 5 May 1987, World Bank, Washington, DC.

Cook, C.C. and Grut, M., 1989, 'Agroforestry in Sub-Saharan Africa: a farmer's perspective', *World Bank Technical Paper* 112, World Bank, Washington, DC.

Coughenour, M.B., Ellis, J.E., Swift, D.M., Coppock, D.L., Gavin, K., McCabe, J.T. and Hart, T.C., 1985, 'Energy extraction and use in a nomadic pastoral ecosystem', *Science* 230: 619–25.

Cousins, B., 1992, *Managing Communal Rangeland in Zimbabwe: Experiences and Lessons*. Commonwealth Secretariat, London.

Cowling, R., 1991, 'Options for rural land use in Southern Africa: an ecological perspective', in M. de Klerk (ed.), *A Harvest of Discontent: The Land Question in South Africa*, IDASA, Cape Town

Croll, E. and Parkin, D. (eds), 1992, *Bush Base, Forest Farm: Culture, Environment and Development*. Routledge, London.

Crosby, A., 1986, *Ecological Imperialism: The Biological Expansion of Europe 900–1900*. Cambridge University Press, Cambridge.

Crummey, D. (ed.), 1986, *Banditry, Rebellion and Social Protest in Africa*. James Currey, London and Heinemann, Portsmouth, NH.

Dankwerts, J.P., 1973, *A Socio-economic Study of Veld Management in the Tribal Areas of Victoria Province*. The Tribal Areas of Rhodesia Research Foundation, Salisbury.

Davies, A.G. and Richards, P., 1991, *Rain Forest in Mende Life: resources and subsistence strategies in rural communities around the Gola North forest reserve (Sierra Leone)*. Report to Economic and Social Committee on Research (ESCOR), UK Overseas Development Administration. University College London.

Davies, S., 1992, 'Green conditionality and food security: winners and losers from the greening of aid', *Journal of International Development* 4 (2): 151–65.

Davies, S (ed.), 1994, 'Knowledge is power? The use and abuse of information in development', *IDS Bulletin* 25 (2).

Davies, S. and Leach, M., 1991, 'Globalism versus villagism: food security and the environment at national and international levels', *IDS Bulletin* 22 (3): 43–50.

Davis, G., 1982, 'Forest reserves', in P.S. Abdu *et al.* (ed.), *Sokoto State in Maps*, University Presses, Ibadan and Oxford, pp. 14–15.

De Ploey, J., 1981, 'The ambivalent effects of some factors of erosion', *Mémoire Institut de Géologie, Universite de Louvain* 31: 171–81.

De Ploey, J., Savat, J. and Moeyersons, J., 1976, 'The differential impact of some soil loss factors on flow, runoff creep and rainwash', *Earth Surface Processes* 1: 151–61.

De Schlippe, P., 1956, *Shifting Cultivation in Africa: The Zande System of Agriculture*. Routledge and Kegan Paul, London.

Dean, W.R.J. and MacDonald, I.A.W., 1994, 'Historical changes in stocking rates of domestic livestock as a measure of semi-arid and arid rangeland degradation in the Cape Province, South Africa', *Journal of Arid Environments* 26: 281–98.

Denny, R.P., Barnes, D.L. and Kennan, T.C.D., 1977, 'Trials of multi-paddock grazing systems on veld 1. An exploratory trial of systems involving 12 paddocks and one herd', *Rhodesian Journal of Agricultural Research* 15: 11–23.

Drinkwater, M., 1989, 'Technical development and peasant impoverishment: land use policy in Zimbabwe's Midlands Province', *Journal of Southern African Studies* 15: 287–305.

Dublin, H., Sinclair, A. and McGlade, J., 1990, 'Elephants and fire as causes of multiple stable states in the Serengeti-Mara woodlands', *Journal of Animal Ecology* 59: 1147–64.

Dunne, T., Dietrich, W.E. and Brunengo, M.J., 1978, 'Recent and past erosion rates in semi-arid Kenya', *Zeitschrift für Geomorphologie* Suppl. Bd. 29: 130–40.

Dupré, G., 1991, 'Les arbres, le fourré et le jardin: les plantes dans la société de Aribinda, Burkina Faso', in G. Dupré (ed.), *Savoirs paysans et développement*. Karthala-ORSTOM, Paris, pp. 181–94.

Durning, A., 1990, 'Apartheid's environmental toll', *Worldwatch Paper* 95, Worldwatch Institute, Washington, DC.

Earle, C., 1988, 'The myth of the southern soil miner: macrohistory, agricultural innovation, and environmental change', in D. Worster (ed.), *The Ends of the Earth: Perspectives on Modern Environmental History*, Cambridge University Press, Cambridge, pp. 175–210.

Egboh, E.E., 1979, 'The establishment of government-controlled forest reserves in Nigeria, 1897–1940', *Savanna* 8 (2): 1–18.

Egboh, E.E., 1985, *Forestry Policy in Nigeria, 1897–1960*. University of Nigeria Press, Nsukka.

Ehrlich, P. and Ehrlich, A., 1990, *The Population Explosion*. Simon and Schuster, New York.

Elliot, J., 1989, 'Soil erosion and conservation in Zimbabwe: political economy and the environment', PhD thesis, Loughborough University.

Ellis, J.E., and Swift, D.M., 1988, 'Stability of African pastoral ecosystems: alternate paradigms and implications for development', *Journal of Range Management* 41: 450–9.

Elwell, H.A., 1983, 'The degrading soil and water resources of the communal areas', *Zimbabwe Science News* 17: 145–7.

Elwell, H.A. and Stocking, M.A., 1988, 'Loss of soil nutrients by sheet erosion is a major hidden farming cost', *Zimbabwe Science News* 22: 79–82.

ENDA/ZERO, 1992, *The Case for Sustainable Development in Zimbabwe: Conceptual Problems, Conflicts and Contradictions*. ENDA/ZERO, Harare.

Fairburn, W.A., 1937, *Investigation of Sylvan Conditions and Land Utilization in Northern Kano and Katsina Provinces.* Northern Nigeria Government, Kaduna.

Fairhead, J., 1992, 'Indigenous technical knowledge and natural resources management in Sub-Saharan Africa: a critical review'. Paper prepared for the Social Science Research Council Project on African Agriculture, Dakar, January.

Fairhead, J. and Leach, M., 1994, 'Contested forests: modern conservation and historical land use in Guinea's Ziama Reserve', *African Affairs* 93 (373): 481–512.

Fairhead, J. and Leach, M., 1996, *Misreading the African Landscape: Society and Ecology in a Forest-savanna Mosaic,* Cambridge University Press, Cambridge and New York.

FAO, 1985, *Tropical Forestry Action Plan.* Food and Agriculture Organization of the United Nations, Committee on Forest Development in the Tropics, Rome.

FAO, 1993, 'Forest resources assessment 1990: Tropical countries', *FAO Forestry Paper* 112, Rome.

FAO/UNDP, 1984, 'Ethiopia: a land resources inventory for land use planning', *Technical Report* 1–6, FAO, Rome.

Ferguson, J., 1990, *The Anti-Politics Machine: 'Development', Depoliticization and Bureaucratic State Power in Lesotho.* Cambridge University Press, Cambridge.

Figueirredo, P., 1986, 'The yield of food crops on terraced and non-terraced land – a field survey of Kenya: report from a minor research task', *Working Paper* 35, Swedish University of Agricultural Sciences, Uppsala.

Fitzgerald, M., 1994, 'Environmental education in Ethiopia: a strategy to reduce vulnerability to famine', in A. Varley (ed.), *Disasters, Development and Environment,* Wiley, Chichester, pp. 125–38.

Fleuret, P., 1985, 'The social organisation of water control in the Taita Hills, Kenya', *American Ethnologist* 12: 103–18.

Ford, J., 1971, *The Role of the Trypanosomiases in African Ecology: A Study of the Tsetse Fly Problem.* Oxford University Press, Oxford.

Fosbrooke, H.A., 1948, 'An administrative survey of the Maasai social system', *Tanganyika Notes and Records* 26: 1–50.

Foucault, M., 1971, 'The order of discourse', in R. Young (ed.), *Untying the Text: A Poststructuralist Reader,* Routledge and Kegan Paul, London.

Foucault, M., 1979, *Discipline and Punish: the Birth of the Prison.* Vintage, New York.

Foucault, M., 1980, *Power/Knowledge: Selected Interviews and Other Writings 1972–1977,* edited by C. Gordon. Harvester Press, Brighton.

Friedel, M.H., 1991, 'Range condition assessment and the concept of thresholds: a viewpoint', *Journal of Range Management* 44: 422–33.

Frost, P., Medina, E., Menaut, J.-C., Solbrig, O., Swift, M. and Walker, B., 1986, 'Responses of savannas to stress and disturbance: a proposal for a collaborative programme of research', *Biology International,* Special Issue 10, IUBS, Paris.

Froude, M., 1974, 'Veld management in the Victoria Province Tribal Areas', *Rhodesia Agriculture Journal* 71: 29–33.

Fruhling, P., 1988, *Utveckling Bättre än Nödhjälp: Om Röda Korsets Katastrofförebyggande arbete i Wollo.* Röda Korset, Stockholm.

Funtowicz, S.O. and Ravetz, J.R., 1992, 'Three types of risk assessment and the emergence of post-normal science', in S. Krimsky and D. Golding (eds), *Social Theories of Risk,* Praeger, Westport, CT, pp. 251–73.

Galaty, J.G. and Bonte, P. (eds), 1991, *Herders, Warriors and Traders: Pastoralism in Africa.* Westview Press, Boulder, CO.

Galaty, J.G., Aronson, D. and Saltzman, P.C. (eds), 1981, 'The Future of Pastoral Peoples'. Proceedings of a conference held in Nairobi, August 1980.

Gambiza, J., 1987, 'Some effects of different stocking intensities on the physical and chemical properties of the soil in a marginal rainfall area of southern Zimbabwe', unpublished MSc thesis, Department of Biological Sciences, University of Zimbabwe.

Gammon, M., 1978, 'A review of experiments comparing systems of grazing management on natural pastures', *Proceedings of the Grasslands Society of Southern Africa* 13: 75–82.

Geiger, R., 1966, *The Climate Near the Ground*. Harvard University Press, Cambridge, MA.

Gellner, E., 1992, *Postmodernism, Reason and Religion*, Routledge, London.

GFA, 1987, *Study on the Economic and Social Determinants of Livestock Production in the Communal areas, Zimbabwe*. Department of Veterinary Services, Harare.

Giddens, A., 1984, *The Constitution of Society: Outline of the Theory of Structuration*. Polity Press, Oxford.

Giddens, A., 1987, *Social Theory and Modern Society*. Basil Blackwell, Oxford.

Glacken, C., 1967, *Traces on the Rhodian Shore: Nature and Culture in Western Thought from Ancient Times to the End of the Eighteenth Century*. University of California Press, Berkeley, CA.

Glantz, M.H. and Orlovsky, N., 1983, 'Desertification: a review of the concept', *Desertification Control Bulletin* 9: 15–22.

Gottlieb, A., 1992, *Under the Kapok Tree: Identity and Difference in Beng Thought*. Indiana University Press, Bloomington and Indianapolis, IN.

Green, W., 1991, 'Lutte contre les feux de brousse', Report for project DERIK, Développement Rural Intégré de Kissidougou, GTZ.

Grove, R.H., 1987, 'Early themes in African conservation: the Cape in the nineteenth century', in D. Anderson and R. Grove (eds), *Conservation in Africa*, Cambridge University Press, Cambridge, pp. 21–39.

Grove, R.H., 1989, 'Scottish missionaries, evangelical discourses and the origins of conservation thinking in southern Africa, 1820–1900', *Journal of Southern African Studies* 15 (2): 163–87.

Grove, R.H., 1990, 'Colonial conservation, ecological hegemony and popular resistance; towards a global synthesis', in J. Mackenzie (ed.), *Imperialism and the Natural World*, Manchester University Press, Manchester, pp. 15–51.

Grove, R.H., 1993, 'Conserving Eden: the European East India Companies and their environmental policies on St Helena, Mauritius and in western India', *Comparative Studies in Society and History* 35: 318–51.

Grove, R.H., 1994, 'Chiefs, boundaries and sacred groves: early nationalism and defeat of colonial conservationism in the Gold Coast and Nigeria, 1870–1915'. Paper presented at the conference on 'Escaping Orthodoxy', Institute of Development Studies, Sussex, 13–14 September.

Grove, R.H., 1995, *Green Imperialism: Colonial Expansion, Tropical Island Edens and the Origins of Environmentalism, 1600–1860*, Cambridge University Press, Cambridge.

Gunn, M.D., 1968, 'Illtyd Buller Pole Evans (1879–1968)', *Bothalia* 10: 131–5.

Hagen, E.E., 1964, *On the Theory of Social Change*. Tavistock, London .

Hair, P.E.H., 1962, 'An account of the Liberian Hinterland c. 1780', *Sierra Leone Studies* 16: 218–26.

Hall, M., 1990, *Farmers, Kings and Traders: The People of Southern Africa 200–1860*, Chicago University Press, Chicago, IL.

Hall, T.D., 1934, 'South African pastures: retrospective and prospective', *South African Journal of Science* XXXI: 59–97.

Hansen, A. 1991, 'The media and the social construction of the environment', *Media, Culture and Society* 13 (4): 443–58.

Hardin, G., 1968, 'The tragedy of the commons', *Science* 162: 1243–8.

Harris, L.D. and Fowler, N.K., 1975, 'Ecosystem analysis and simulation of the Mkomazi Reserve, Tanzania', *East African Wildlife Journal* 13: 325–46.

Hayes, J.J., 1986, 'Not enough wood for the women: how modernization limits access to resources in the domestic economy of rural Kenya', unpublished MA thesis, Clark University, Worcester, MA.

Haylett, D., Murray, C. and Eriksson, F., 1932, 'Preliminary notes on intensity of grazing experiment', *Rhodesian Agriculture Journal* 29: 641–5.

Helldén, U., 1991, 'Desertification: time for an assessment?', *Ambio* 20 (8): 372–83.

Henning, R.O., 1941, 'The furrow-makers of Kenya', *Geographical Magazine* 12: 268–79.

Herskovits, M., 1926, 'The cattle complex in East Africa', *American Anthropologist* 28: 230–72, 361–80, 494–528, 630–64.

Herweg, K., 1992, 'Major constraints to effective soil conservation. Experiences in Ethiopia', Paper presented to the Seventh International Soil Conservation Conference: People Protecting their Soil, 27–30 September, ISCO, Sidney.

Herweg, K., n.d., 'Problems of acceptance and adaptation of soil conservation in Ethiopia', unpublished paper, Soil Conservation Research Project, Addis Ababa.

Herweg, K. and Grunder, M., 1991, 'Soil conservation research project, 9, Eighth Project report, Year 1988', Soil Conservation Research Project, Berne.

Heyer, J.U., 1966, 'Agricultural development and peasant farming in Kenya', unpublished PhD thesis, University of London.

Hirschman, A.O.,1968, *Development Projects Observed.* Brookings Institution, Washington, DC.

Hoben, A., 1972, *Land Tenure Among the Amhara of Ethiopia.* Chicago University Press, Chicago, IL.

Hoben, A., 1975, 'Family, land, and class in northwest Europe and northern highland Ethiopia', in H.G. Marcus (ed.), *Proceedings of the First United States Conference on Ethiopian Studies, 1973,* African Studies Center, Michigan State University, East Lansing, MI, pp. 157–70.

Hoben, A., 1980, 'Agricultural decision making in foreign assistance: an anthropological analysis', in P. Barlett (ed.), *Agricultural Decision Making: Anthropological Contributions to Rural Development,* Academic Press, Orlando, FL, pp. 337–69.

Hoben, A., 1989, 'USAID: Organizational and institutional issues and effectiveness', in R.J. Berg and D.F. Gordon (eds), *Cooperation for International Development: the United States and the Third World in the 1990s,* Lynne Rienner, Boulder, CO, pp. 253–78.

Hoben, A., 1994, 'The cultural and political construction of environmental policy in Africa: some theoretical considerations', Paper presented at the SSRC Workshop on the Dynamics of Production and Transmission of Development Ideas, 13–15 May, University of Michigan, Ann Arbor, MI.

Hoben, A., 1995, 'Paradigms and politics: the cultural construction of environmental policy in Ethiopia'. *World Development* 23 (6): 1007–22 .

Hoffmann, M.T., 1993, 'The potential value of historical ecology to environmental monitoring', in C. Marais and D.M. Richardson (eds), *Monitoring Requirements for Fynbos Management,* Foundation for Research Development, FRD Programme Report Series no.11.

Hoffmann, M.T. and Cowling, R.M., 1990, 'Vegetation change in the semi-arid eastern Karoo over the last 200 years: and expanding Karoo fact or fiction?', *South African Journal of Science* 86: 286–94.

Hoffmann, M.T., Bond, W.J. and Stock, W.D., 1995, 'Desertification of the Eastern Karoo, South Africa: conflicting palaeoelogical, historical and soil isotopic evidence', in *Environmental Monitoring and Assessment.*

Holling, C.S., 1973, 'Resilience and stability of ecological systems', *Annual Review of Ecology and Systematics* 4: 1–23.

Holmberg, G., 1990, 'An economic evaluation of soil conservation in Kitui District Kenya', in J.A. Dixon, D.E. James and P.B. Sherman (eds), *Dryland Management: Economic Case Studies*, Earthscan, London.

Holmgren, E. and Johansson, G., 1987, 'Comparisons between terraced and non-terraced land in Machakos District, Kenya', *Machakos Report 1987*, Soil and Water Conservation Branch, Ministry of Agriculture, Nairobi.

Homewood, K.M., 1992a, 'Development and the ecology of Maasai pastoralist food and nutrition', *Ecology of Food and Nutrition* 29: 61–80.

Homewood, K.M., 1992b, 'Patch production by cattle', *Nature* 359: 109–10.

Homewood, K.M., 1994, 'Pastoralists, environment and development in East African rangelands', in B. Zaba and J. Clarke (eds), *Environment and Population Change*. Ordina editions, Paris.

Homewood, K.M. and Rodgers, W.A., 1984, 'Pastoralism and conservation', *Human Ecology* 12: 431–41.

Homewood, K.M. and Rodgers, W.A., 1987, 'Pastoralism, conservation and the overgrazing controversy', in D.M. Anderson and R. Grove (eds), *Conservation in Africa: People, policies and practice*. Cambridge University Press, Cambridge, pp. 111–28.

Homewood, K.M. and Rodgers, W.A., 1991, *Maasailand Ecology: Pastoralist Development and Wildlife Conservation in Ngorongoro, Tanzania*. Cambridge University Press, Cambridge.

Horsman, J., 1975, 'Production forestry in the Zaria area', *Savanna* 4 (1): 70–4.

Howard, W.J., 1976, 'Land Resources of Central Nigeria', *Land Resource Report* 9, Land Resources Division, Ministry of Overseas Development, Surbiton.

Hudock, A.C. 1995. 'Sustaining Southern NGOs in resource-dependent environments', *Journal of International Development* 7(4): 653–68.

Hudson, N.W., 1957, 'Erosion control research. Progress report on experiments at Henderson Research Station, 1953–56', *Rhodesia–Agricultural Journal* 54: 297–323.

Hultin, J., 1988, *Farmers' Participation in the Wollo Program*, Swedish International Development Authority (SIDA), Stockholm.

Huntley, B., Siegfried, R. and Sunter, C., 1989, *South African Environments into the 21st Century*. Tafelberg, Cape Town.

Husband, A.D. and Taylor, A.P., 1931, 'Studies on the improvement of natural veld pastures', *Rhodesian Agricultural Journal* 28: 154–69.

Hyman, E.L., 1993, 'Forestry policies and programmes for fuelwood supply in Northern Nigeria', *Land Use Policy* 23 (1): 26–43.

Ibrahim, F.N., 1984, 'Ecological imbalance in the Republic of the Sudan - with reference to desertification in Darfur', *Bayreuther Geowissenschaftliche Arbeiten* 6. Druckhaus Bayreuth Verlagsgesellschaft, Bayreuth.

IIED, 1994, 'Whose Eden? An overview of community approaches to wildlife management', Report to Overseas Development Administration. International Institute for Environment and Development, London.

ILCA (International Livestock Centre for Africa), 1980, 'Pastoral development projects', *ILCA Bulletin* 9, Nairobi.

IUCN (International Union for Conservation of Nature and Natural Resources), 1990, *Ethiopia National Conservation Strategy*, Phase I Report, prepared for the Government of the People's Democratic Republic of Ethiopia with the assistance of IUCN, Addis Ababa.

Jaffee, S.M., 1994, 'Contract farming in the shadow of competitive markets: the experience of Kenyan Horticulture', in P.D. Little and M.J. Watts (eds), *Living Under Contract: Contract Farming and Agrarian Transformation in Sub-Saharan Africa*, University of Wisconsin Press, Madison, WI.

Jeffreys, M.W., 1950, 'Feux de brousse', *Bulletin IFAN* 3: 682–7100.

Jones, B., 1938, 'Dessication and the West African colonies', *Geographical Journal* 88 (5): 401–23.

Kano State Afforestation Committee (KSAC), 1987, *Report on the Need for Mass Tree Planting in Kano State*. Ministry of Agriculture and Natural Resources, Kano.

Kano State Agricultural and Rural Development Authority (KNARDA), n.d., *Guide to Tree Planting and Protection Programme. Handbook for Extension Staff*. KNARDA, Kano.

Kano State Committee on Alternative Sources of Energy (KSCASE), 1989, *Report on Alternative Sources of Energy*. Ministry of Agriculture and Natural Resources, Kano.

Keay, R.W.J., 1959, 'Derived savanna – derived from what?', *Bulletin IFAN*, Series A 2: 427–38.

Kenya, Government of, 1962, *African Land Development in Kenya*, 1946–62. Ministry of Agriculture, Animal Husbandry and Water Resources, Nairobi.

Kerr, G.R.G., 1940, 'Some notes on the forestry situation in Northern Nigeria', *Nigerian Forester* 1 (1): 20–2.

Kimmage, K., 1990, 'NEAZDP socio-economic survey. Summary results', Mimeo, NEAZDP, Gashua.

Kiome R. and Stocking, M.A., 1993, 'Soil conservation in semi-arid Kenya', *Bulletin* 61, Natural Resources Institute, Chatham.

Kiome, R. and Stocking, M.A., 1995, 'Rationality of farmer perception of soil erosion: the effectiveness of soil conservation in semi-arid Kenya', *Global Environmental Change*.

Kipkorir, B.E., 1983, 'Historical perspectives on development of the Kerio Valley', in B.E. Kipkorir, R.C. Soper and J.W. Ssennnyonga (eds), *Kerio Valley: Past, Present and Future*, Institute of African Studies, University of Nairobi, pp. 1–11.

Kuhn, T., 1962, *The Structure of Scientific Revolutions*. Chicago University Press, Chicago, IL.

Lacey, C. and Longman, D., 1993, 'The press and public access to the environment and development debate', *The Sociological Review* 41(2): 207–43.

Lal, R., 1976, 'Soil erosion on alfisols in Western Nigeria. IV. Nutrient element losses in runoff and eroded sediments', *Geoderma* 16: 403–17.

Lal, R., 1987, *Tropical Ecology and Physical Edaphology*. Wiley, Chichester.

Lamprey, H.F., 1975, 'Report of the desert encroachment reconnaissance in Northern Sudan: 21 October to 10 November 1975', typescript. UN Environment Programme - UNESCO, Nairobi.

Lane, C., 1991, 'Alienation of Barabaig pastureland: policy implications for pastoral development in Tanzania', *Drylands Programme Report,* International Institute for Environment and Development (IIED), London.

Lane, C. and Swift, J., 1989, 'East African pastoralism: common land, common problems', *IIED Drylands Issues Paper* 8, International Institute for Environment and Development (IIED), London.

Lane, C., Moorehead, R., Mayers, J., Eaton, D., Dalal-Clayton, B., Koziell, I. and Jennings, S., 1994, *Whose Eden? An Overview of Community Approaches to Wildlife Management*. International Institute for Environment and Development (IIED), London.

Laycock, W.A., 1991, 'Stable states and thresholds of range condition on North American rangelands: a viewpoint', *Journal of Range Management* 44: 427–73.

Leach, G. and Mearns, R., 1988, *Beyond the Woodfuel Crisis: People, Land and Trees in Africa*. Earthscan Publications, London.

Leach, M., 1994, *Rainforest Relations: Gender and Resource Use Among the Mende of Gola, Sierra Leone*. Edinburgh University Press, Edinburgh, for the International African Institute.

Leach, M. and Fairhead, J., 1993, 'Whose social forestry and why? People, trees and managed continuity in Guinea's forest-savanna mosaic'. *Zeitschrift für Wirtschaftsgeographie* 37(2): 86–101.

Leach, M. and Fairhead, J., 1994, 'The forest islands of Kissidougou: social dynamics of environmental change in West Africa's forest-savanna mosaic', report to Economic and Social Committee on Research of the Overseas Development Administration. Institute of Development Studies, University of Sussex.

Leach, M. and Fairhead, J., 1995, 'Ruined settlements and new gardens: gender and soil ripening among Kuranko farmers in the forest-savanna transition zone', *IDS Bulletin* 26 (1): 24–32.

Leach, M. and Mearns, R., 1991, 'Editorial' and 'Challenges for social science research' in 'Environmental Change, Development Challenges', *IDS Bulletin* 22 (4): 1–4, 50–52.

Leiss, W., 1972, *The Domination of Nature*. George Braziller, New York.

Lindblom, K.G., 1920, *The Akamba of British East Africa*. Appelborgs Boktrycheri Aktieborg, Uppsala.

Lindsay, W.K., 1987, 'Integrating parks and pastoralists: some lessons from Amboseli', in D.M. Anderson and R. Grove (eds), *Conservation in Africa: People, Policies and Practice*. Cambridge University Press, Cambridge, pp. 149–67.

Little, P.D., 1992, *The Elusive Granary: Herder, Farmer and the State in Northern Kenya*. Cambridge University Press, Cambridge.

Lockwood, M., 1991, 'Farmers' perceptions of population pressure in southern Kano', mimeo, African Studies Centre, Cambridge University.

Long, N. (ed.), 1989, 'Encounters at the interface: a perspective on social discontinuities in rural development', *Wageningse Sociologische Studies* 27, Wageningen Agricultural University, The Netherlands.

Long, N. and Long, A. (eds) 1992, *Battlefields of Knowledge: The Interlocking of Theory and Practice in Social Research and Development*. Routledge, London.

Long, N., and Van der Ploeg, J., 1989, 'Demythologizing planned intervention: an actor perspective', *Sociologia Ruralis* xxix (3/4): 227–49.

Lowdermilk, W.C., 1935, 'Civilization and soil conservation', *Rhodesian Agriculture Journal* 32, 553–7.

Lowe, P. and Morrison, D., 1984, 'Bad news or goods news: environmental politics and the mass media', *The Sociological Review* 32 (1): 75–90.

Ludlum, Governor, 1808, 'To the editor of the Sierra Leone Gazette', *The Sierra Leone Gazette*, April 1808: 16.

Lugard, F., 1970, *Political Memoranda: Revision of Instructions to Political Officers on Subjects Chiefly Political and Administrative, 1913–1919*, edited by A.H.M. Kirk-Greene, Frank Cass, London.

Lundgren, L., Taylor, G. and Ingevall, A., 1993, *From Soil Conservation to Land Husbandry: Guidelines Based Upon SIDA's Experience*. Natural Resources Management Division, Swedish International Development Authority, Stockholm.

Magadza, C.H.D., 1992, *Conservation for Preserving the Environment*. Mambo Press, Gweru.

Matingu, M.N., 1974, 'Rural to rural migration and employment: a case study in a selected area of Kenya', unpublished MA thesis, University of Nairobi.

Matose, F., 1991, 'Villagers as woodland managers', mimeo, Forestry Research Centre, Harare.

Matose, F. and Mukamuri, B., 1994, 'Trees, people and communities in Zimbabwe's communal lands', in I. Scoones and J. Thompson (eds), *Beyond Farmer First: Rural People's Knowledge, Agricultural Research and Extension Practice*, Intermediate Technology Publications, London, pp. 69–74.

May, R., 1973, *Stability and Complexity in Model Ecosystems*. Princeton University Press, Princeton, NJ.

May, R., 1977, 'Thresholds and breakpoints in ecoystems with a multiplicity of stable states', *Nature* 269: 471–7.

Mbithi, P. and Barnes, C., 1975, *The Spontaneous Settlement Problem in Kenya*. East African Literature Bureau, Nairobi.

McCabe, J.T., 1990, 'Turkana pastoralism: a case against the tragedy of the commons', *Human Ecology* 18 (1): 81–103.

McCann, J., 1987, *From Poverty to Famine in Northeast Ethiopia: A Rural History 1900–1935*. University of Pennsylvania Press, Philadelphia, PA.

McCann, J., 1995, *People of the Plow: An Agricultural History of Ethiopia, 1800–1990*. University of Wisconsin Press, Madison, WI.

McGregor, J., 1991, 'Woodland resources: ecology, policy and ideology. An historical case study of woodland use in Shurugwi communal area, Zimbabwe', unpublished PhD thesis, Loughborough University.

McIntosh, R., 1987, 'Pluralism in ecology', *Annual Review of Ecological Systematics* 18: 321–41.

McNeill, W.H., 1976, *Plagues and Peoples*. Penguin, London.

Mduma, S.R., 1988, 'Mkomazi Game Reserve: dangers and recommended measures for its survival', *Miombo* 1: 17–19.

Mearns, R., 1991, 'Structural adjustment and the environment: reflections on scientific method', *IDS Discussion Paper* 284. Institute of Development Studies, University of Sussex.

Mearns, R., 1995, 'Institutions and natural resource management: access to and control over woodfuel in East Africa', in T. Binns (ed.), *People and Environment in Africa*, Wiley, Chichester, pp. 103–14.

Mesfin W.-M., 1991, *Suffering under God's Environment: A Vertical Study of the Predicament of Peasants in North-central Ethiopia*, African Mountains Association, Geographica Bernensia, University of Berne, Switzerland.

Meyers, L.R., 1981, 'Organization and administration of integrated rural development in semi-arid areas: the Machakos Integrated Development Program', report prepared for the Office of Rural Development and Development Administration, Development Support Bureau, Agency for International Development. Contract No. AID/DSAN-C-0212. Washington, DC.

Meyers, L.R., 1982, 'Socio-economic determinants of credit adoption in a semi-arid district of Kenya', unpublished PhD dissertation, Cornell University.

Mikhebi, A.W., Knipscheer, H.C. and Sullivan, G., 1991, 'The impact of foodcrop production on sustained livestock production in semi-arid regions of Kenya', *Agricultural Systems* 35: 339–51.

Millimouno, D., 1993, 'La gestion locale du feu dans la préfecture de Kissidougou', *COLA Working Paper 5*, Institute of Development Studies, University of Sussex.

Millington, A.C., 1987, 'Environmental degradation, soil conservation and agricultural policies in Sierra Leone, 1895–1984', in D.M. Anderson and R. Grove (eds), *Conservation in Africa: People, Policies and Practice*, Cambridge University Press, Cambridge, pp. 229–48.

Milton, K., 1993, 'Introduction: environmentalism and anthropology', in K. Milton (ed.), *Environmentalism: The View from Anthropology*. Routledge, London, pp. 1–17.

Milton, S.J. and Hoffman, M.T., 1994, 'The application of state-and-transition models to rangeland research and management in arid succulent and semi-arid grassy Karoo, South Africa', *African Journal of Range and Forestry Science* 11 (1): 18–26.

Moeyersons, J., 1978, 'The behaviour of stones and stone implements buried in consolidating and creeping Kalahari sands', *Earth Surface Processes* 3: 115–28.

Moore, H.L., 1986, *Space, Text and Gender: an Anthropological Study of the Marakwet of Kenya.* Cambridge University Press, Cambridge.

Morgan, W.T.W., 1985, *Nigeria.* Longman, Harlow, Essex.

Mortimore, M. J., 1972, 'Some aspects of rural-urban relations in Kano, Nigeria', in P. Vennetier (ed.), *La croissance urbaine en Afrique noire et à Madagascar,* Centre Nationale de la Recherche Scientifique, Paris, pp. 874–9.

Mortimore, M., 1989, *Adapting to Drought: Farmers, Famines and Desertification in West Africa.* Cambridge University Press, Cambridge.

Mortimore, M.J., Essiet, E.U. and Patrick, S.P., 1990, *The Nature, Rate and Effective Limits of Intensification in the Smallholder Farming System of the Kano Close-Settled Zone.* Federal Agricultural Coordinating Unit, Ibadan.

Mortimore, M., Tiffen, M. and Gichuki, F., 1993, 'Sustainable growth in Machakos', *ILEIA Newsletter* 4/93: 6–10.

Moss, R.P., 1982, 'Reflections on the relation between forest and savanna in Tropical West Africa', *University of Salford Discussion Papers in Geography* 23, Department of Geography, University of Salford.

Moss, R.P. and Morgan, W., 1977, 'Soils, plants and farmers in West Africa: Parts 1 & 2', in J.P. Garlick and R.W.J. Keay (eds), *Human Ecology in the Tropics,* Symposia of the Society for the Study of Human Biology, Vol. 16, Taylor and Francis, London, pp. 27–77.

Mtetwa, R., 1978, 'Myth or reality: the "cattle complex" in S.E. Africa, with special reference to Rhodesia', *Zambezia* VI: 23–35.

Mukamuri, B., 1988, 'Rural environmental conservation strategies in south-central Zimbabwe: an attempt to describe Karanga thought patterns, perceptions and environmental control', Paper presented at the African Studies Association conference, Cambridge, September.

Mukhebi, A.W., Knipscheer, H.C. and Sullivan, G., 1991, 'The impact of foodcrop production on sustained livestock production in semi-arid regions of Kenya', *Agricultural Systems* 35: 339–51.

Munro, J.F., 1975, *Colonial Rule and the Kamba: Social Change in the Kenya Highlands 1889–1939.* Clarendon Press, Oxford.

Murombedzi, J., 1992, 'Decentralising common property resources management: a case study of Nyaminyami District Council'. *Drylands Network Programme Issues Paper* 30, International Institute for Environment and Development (IIED), London.

Murphy, W., 1990, 'Creating the appearance of consensus in Mende political discourse', *American Anthropologist* 92 (1): 24–41.

Mustafa, K., 1993, 'Eviction of pastoralists from Mkomazi Game Reserve in Tanzania: a statement', mimeo, International Institute for Environment and Development (IIED), London.

Mwalyosi, R.B., 1992, 'Influence of livestock grazing on range condition in south-west Maasailand, Northern Tanzania', *Journal of Applied Ecology* 29: 581–8.

Nelson, R., 1988, 'Dryland management: the "desertification" problem', *Environment Department Working Paper* 8. World Bank, Washington, DC.

Nhira, C., 1992, 'Integrated natural resource management in communal areas: problems and opportunities in Kanyati communal area', in B. Cousins (ed.), *Institutional Dynamics in Communal Grazing Regimes in Southern Africa,* CASS, Harare, pp. 125–38.

Niamir, M., 1990, 'Herders' decision-making in natural resources management in arid and semi-arid Africa', *Community Forestry Note* 4, Food and Agricultural Organization, Rome.

Nicholson, S.E., 1979, 'The methodology of historical climate reconstruction and its application to Africa', *Journal of African History* 20 (1): 31–49.

Nigerian Environmental Study/Action Team (NEST), 1991, *Nigeria's Threatened Environment: A National Profile,* NEST, Ibadan.

Noble, J.C., 1986, 'Plant population ecology and colonal growth in arid rangeland ecosystems', in P.J. Joss, P.W. Lynch, and O.B. Williams (eds), *Rangelands: a Resource Under Siege*, Proceedings of the Second International Rangeland Congress, Cambridge University Press, Cambridge .

O'Connor, T.G., 1985, 'A synthesis of field experiments concerning the grass layer in the savanna regions of southern Africa', *South African National Scientific Programme Report* 114, CSIR, Pretoria.

Oluwasanmi, H.A., 1966, *Agriculture and Nigerian Economic Development*, Oxford University Press, Ibadan.

Östberg, W., 1991, '"Land is coming up": Burungee throughts on soil erosion and soil formation', *EDSU Working Paper* 11, School of Geography, University of Stockholm.

Ostrom, E., 1990, *Governing the Commons: The Evolution of Institutions for Collective Action*, Cambridge University Press, Cambridge.

Pankhurst, H., 1992, *Gender, Development and Identity: An Ethiopian Study.* Zed Press, London.

Pankhurst, R. and Johnson, D.H.. 1988, 'The great drought and famine of 1888–92 in northeast Africa', in D.H. Johnson and D.M. Anderson (eds), *The Ecology of Survival: Case Studies from Northeast African History*, Lester Cook Academic Publishing, London.

Peberdy, J.R., 1958, *Machakos District Gazetteer.* Department of Agriculture, Machakos District Office, Kenya.

Peden, D.G., 1987, 'Livestock and wildlife population distributions in relation to aridity and human population in Kenya', *Journal of Range Management* 40: 67–71.

Peet, R. and Watts, M., 1993, 'Introduction: development theory and environment in an age of market triumphalism', in theme issue on 'Environment and Development', *Economic Geography* 69 (3): 227–53.

Pellew, R.A.P., 1983, 'The impacts of elephant, giraffe and fire upon the *Acacia tortilis* woodlands of the Serengeti', *African Journal of Ecology* 21: 41–74.

Penwill, D.J., 1951, *Kamba Customary Law: Notes Taken in the Machakos District of Kenya Colony.* East African Literature Bureau, Nairobi, and Macmillan, London.

Phillips, J., 1934, 'Succession, development, the climax and the complex organism: an analysis of concepts', *Journal of Ecology* 12: 554–71.

Phillips, J., 1938, 'Deterioriation in the vegetation of the Union of South Africa', *South African Journal of Science* 35: 476–84.

Phimister, I., 1978, 'Meat and monopolies: beef cattle in southern Rhodesia, 1890–1928', *Journal of African History* 19: 391–414.

Pimentel. D. *et al.*, 1995, 'Environmental and economic costs of soil erosion and conservation benefits', *Science* 267: 1117–23.

du Plessis, E., 1987, 'Obituary: John Frederick Vicars Phillips (1899–1987)', *Bothalia* 17: 267–8.

Pole-Evans, I., 1920, 'The veld: its resources and dangers', *South African Journal of Science* 17: 1–34.

Pole-Evans, I., 1932, 'Pastures and their management', *Rhodesian Agriculture Journal* 29: 912–920.

Pole-Evans, I., 1939a, 'Pasture research in South Africa', *Progress Report* 2. Government Printer, Pretoria.

Pole-Evans, I., 1939b, *Report of a Visit to Kenya.* Government Printer, Nairobi.

Ponsart-Dureau, M-C., 1986, 'Le pays Kissi de Guinée forestière: contribution à la connaissance du milieu; problématique de développement', Mémoire, Ecole Supérieure d'Agronomie Tropicale, Montpellier.

Poore, M.E.D. and Fries, C., 1986, 'The ecological effects of eucalyptus', *FAO Forestry Paper* 59, Rome.

Potkanski, T., forthcoming, 'Property concepts, herding patterns and management of natural resources among the Ngorongoro and Salei Maasai of Tanzania', IIED, London.

Pretty, J., 1994, 'Alternative systems of inquiry for a sustainable agriculture', *IDS Bulletin* 25 (2): 37–48.

Pullan, R.A., 1974, 'Farmed parkland in West Africa', *Savanna* 3 (2): 119–51.

Rahmato, D., 1985, *Agrarian Reform in Ethiopia.* Red Sea Press, Trenton, New Jersey.

Rahmato, D., 1987, *Famine and Survival Strategies: A Case Study from Northeast Ethiopia.* Institute of Development Research, Addis Ababa University, Addis Ababa.

Rahmato, D., 1991, 'Agrarian change and agrarian crisis: state and peasantry in post-revolution Ethiopia', mimeo, Institute of Development Research, Addis Ababa University, Addis Ababa.

Ramphele, M., ed., 1991, *Restoring the Land: Environment and Change in Post-Apartheid South Africa.* Panos, London.

Ranesford, O., 1983, *Bid the Sickness Cease: Disease in the History of Black Africa.* John Murray, London.

Ranger, T., 1985, *Peasant Consciousness and Guerilla War in Zimbabwe.* James Currey, London.

Redclift, M., 1993, 'Sustainable development and popular participation: a framework for analysis', in D. Ghai and J.M. Vivian (eds), *Grassroots Environmental Action: People's Participation in Sustainable Development.* Routledge, London, pp. 23–49.

République de Guinée, 1988, *Politique forestière et plan d'action.* Plan d'Action Forestier Tropical 1988, Conakry.

Rhodes, S.L., 1991, 'Rethinking desertification: what do we know and what have we learned?' *World Development* 19 (9): 1137–43.

Richards, P., 1985, *Indigenous Agricultural Revolution: Ecology and Food Production in West Africa.* Allen and Unwin, Hemel Hempstead.

Rivière, C., 1971, *Mutations sociales en Guinée.* Editions Marcel Rivière et Cie, Paris.

Robertson, G., 1987, 'Plant dynamics' in G. Caughley, N. Shepherd and J. Short (eds), *Kangaroos: Their Ecology and Management in the Sheep Rangelands of Australia,* Cambridge University Press, Cambridge.

Robins, S., 1994, 'Contesting the social geometry of bureaucratic state power: a case study of land-use planning in Matabeleland, Zimbabwe', *Social Dynamics,* June, University of Cape Town.

Roe, E., 1991, '"Development narratives" or making the best of blueprint development', *World Development* 19 (4): 287–300.

Roe, E., 1994, 'New framework for an old tragedy of the commons and an aging common property resource management', *Agriculture and Human Values* 11 (1): 29–36.

Roe, E., 1995, 'Except-Africa: postscript to a special section on development narratives', *World Development* 23 (6): 1065–70.

Romyn, A., 1932, 'The possibility of cultivated pastures', *Rhodesian Agricultural Journal* 29: 1125–29.

Romyn, A., 1935, 'Cattle improvement and cattle breeding policy in southern Rhodesia', *Rhodesia Agriculture Journal* 32: 98–107.

Roose, E.J., 1975, *Erosion et Ruisellement en Afrique de L'ouest. Vingt Années de Mesures en Petites Parcelles Experimentales.* ORSTOM, Abidjan.

Roux, E.C., 1946, *The Veld and the Future: A Book on Soil Erosion for South Africans.* African Bookman, South Africa.

Roux, P.W. and Vorster, M., 1983, 'Vegetation change in the Karoo', *Proceedings of the Grassland Society of Southern Africa* 16: 25–9.

Roux, P.W, Vorster, M., Zeeman, P.J.L. and Wentzel, D., 1981, 'Stock production in the Karoo region', *Proceedings of the Grassland Society of Southern Africa* 16: 29–35

Royal Geographical Society, 1994, *Scientific Report for Mkomazi Ecological Research Programme 1994–1996*, RGS, London.

Rukandema, M., Mavua, J.K. and Audi, P.O., 1981, 'The farming system of lowland Machakos, Kenya: farm survey results from Mwala', Farming Systems Economic Research Programme *Technical Report* (Kenya) 1, Ministry of Agriculture, Nairobi.

Sagua, V.O., Ojanuga, A.U., Anabor, E.E., Kio, P.R.O., Kalu, A.E. and Mortimore, M.J. (eds), 1987, *Ecological Disasters in Nigeria: Drought and Desertification.* Federal Ministry of Science and Technology, Lagos.

Sambrook, F. and Gisherford, W., 1935, 'Cattle and the meat industry in southern Rhodesia', *Rhodesian Agricultural Journal* 33: 853–5.

Sandford, S., 1982a, 'Livestock in the communal areas of Zimbabwe', Report prepared for the Ministry of Lands, Rural Resettlement and Rural Development, Harare.

Sandford, S., 1982b, 'Pastoral strategies and desertification: opportunism and conservatism in dry lands', in B. Spooner and H.S. Mann (eds), *Desertification and Development: Dryland Ecology in Social Perspective.* Academic Press, London.

Sandford, S., 1983, *Management of Pastoral Development in the Third World.* Wiley, Chichester.

Savory, A., 1988, *Holistic Resource Management.* Island Press, Washington, DC.

Schnell, R., 1952, 'Contribution à une étude phyto-sociologique et phyto-géographique de l'Afrique Occidentale: les groupements et les unités géo-botaniques de la région guinéenne', Dakar, *Mémoires de l'IFAN* no. 18, Mélanges botanique: 41–236.

Scholes, R.J. and Walker, B.H., 1993, *An African Savanna: Synthesis of the Nylsvley study.* Cambridge University Press, Cambridge.

Scoones, I., 1989a, 'Economic and ecological carrying capacity: implications for livestock development in Zimbabwe's communal areas', *ODI Pastoral Development Network Paper* 27b. Overseas Development Institute, London.

Scoones, I., 1989b, 'Patch use by cattle in dryland Zimbabwe; farmer knowledge and ecological theory', *ODI Pastoral Development Network Paper* 28b. Overseas Development Institute, London.

Scoones, I., 1990, 'Livestock populations and the household economy: a case study from southern Zimbabwe', unpublished PhD thesis, University of London.

Scoones, I., 1992, 'The economic value of livestock in the communal areas of southern Zimbabwe', *Agricultural Systems* 39: 339–59.

Scoones, I. (ed.), 1995, *Living with Uncertainty: New Directions in Pastoral Development in Africa.* Intermediate Technology Publications, London.

Scoones, I. and Cousins, B., 1994, 'Struggle for control over wetland resources in Zimbabwe', *Society and Natural Resources* 7: 579–94.

Scoones, I. and Matose, F., 1993, 'Local woodland management: constraints and opportunities for sustainable resource use', in P. Bradley. and K. McNamara, (eds) *Living with Trees: Policies for Forestry Management in Zimbabwe.* World Bank Technical Paper, 210, World Bank, Washington, DC.

Scoones, I and Thompson, J. (eds), 1994, *Beyond Farmer First: Rural People's Knowledge, Agricultural Research, and Extension Practice.* Intermediate Technology Publications, London.

Scott, J., 1985, *Weapons of the Weak: Everyday Forms of Peasant Resistance.* Yale University Press, New Haven, CT.

Screenivas, L., Johnston, J.R. and Hill, H.W., 1947, 'Some relationships of vegetarian and soil detachment in the erosion process', *Proceedings of the Soil Science Society of America* 12: 471–4.

Sears, P.B., 1935, *Deserts on the March*. First published University of Oklahoma Press, Norman

Sembony, G., 1988, 'Mkomazi shall not die', *Malihai Clubs Newsletter*, Dar es Salaam.

Sessay, M.F. and Stocking, M.A., 1993, 'Soil productivity and fertility maintenance of a degraded oxisol in Sierra Leone', Paper presented at SCOPE Workshop on Sustainable Land Management in Semi-arid and Sub-humid Zones, Dakar, Senegal, 15–19 November.

Seymour, G. L., 1860, 'The journal of the journey of George L. Seymour to the interior of Liberia: 1858', *New York Colonization Journal* 105, 108, 109, 111, 112.

Shaw, J., 1873, 'On some of the changes going on in the South-African vegetation through the introduction of the Merino sheep', Report of the British Association for the Advancement of Science, 43rd meeting, Transactions of the Sections, 105.

Shaxson, T.F., Hudson, N.W., Sanders, D.W., Roose, E. and Moldenhauer, W.C., 1989, *Land Husbandry: A Framework for Soil and Water Conservation*. Soil and Water Conservation Society, Ankeny, IA.

Shepherd, G., 1989, 'The reality of the commons: answering Hardin from Somalia', *Development Policy Review* 7: 51–63.

Shepherd, G., 1992, *Managing Africa's Tropical Dry Forests: A Review of Indigenous Methods*. Overseas Development Institute, London.

Shipton, P., 1989, 'Land and the limits of individualism: population growth and tenure reform south of the Sahara', *HIID Development Discussion Paper* 320, Harvard Institute for International Development, Cambridge, MA.

Showers, K., 1989, 'Soil erosion in the Kingdom of Lesotho: origins and colonial response, 1830s-1950s', *Journal of Southern African Studies* 15 (2): 263–86.

Silviconsult Ltd., 1991, *Northern Nigeria Household Energy Study*. Consultancy report to the Federal Forestry Management Evaluation and Coordinating Unit, Ibadan, Nigeria. Silviconsult, Bjared, Sweden.

Sinclair, A.R.E. and Fryxell, J.M., 1985, 'The Sahel of Africa: ecology of a disaster', *Canadian Journal of Zoology* 63: 987–94.

Smith, E.L., 1988, 'Successional concepts in relation to range condition assessment', in P.T. Tueller (ed.), *Vegetation Science Applications for Rangeland Analysis and Management*. Kluuer Academic Publishers, Dordrecht.

Sobania, N.W., 1988, 'Pastoral migration and colonial policy: a case study from northern Kenya', in D.H. Johnson and D.M. Anderson (eds), *The Ecology of Survival: Case Studies from Northeast African History*, Lester Cook Academic Publishing, London.

Sobania, N.W., 1990, 'Social relationships as an aspect of property rights: Northern Kenya in the pre-colonial and colonial periods', in P.T.W. Baxter and R. Hogg (eds), *Property, Poverty and People: Changing Property Rights and Problems of Pastoral Development*, Department of Social Anthropology and International Development Centre, Manchester University.

Soper, R., 1983, 'A survey of the irrigation systems of the Marakwet', in B.E. Kipkorir, R. Soper and J.W. Ssennyonga (eds), *Kerio Valley: Past, Present and Future*. Institute of African Studies, University of Nairobi, pp. 75–9.

Southern Rhodesia, 1939, 'Report of the Commission to enquire into the preservation of the natural resources of the colony', (McIllwaine report), Government Printers, Salisbury.

Sprugel, D.G., 1991, 'Disturbance, equilibrium, and environmental variability: what is "natural" vegetation in a changing environment?', *Biological Conservation* 58: 1–18.

Ssennyonga, J.W., 1983, 'The Marakwet irrigation system as a model of a systems approach to water management', in B.E. Kipkorir, R. Soper and J.W.

Ssennyonga (eds), *Kerio Valley: Past, Present and Future*, Institute of African Studies, University of Nairobi, pp. 96–111.

Ståhl, M., 1990 'Constraints to environmental rehabilitation through people's participation in the northern Ethiopian highlands', United Nations Research Institute for Social Development, *Discussion Paper* 13, UNRISD, Geneva.

Stamp, L.D., 1940, 'The southern margin of the Sahara: comments on some recent studies on the question of desiccation in West Africa', *Geographical Review* 30: 297–300.

Stebbing, E.P., 1935, 'The encroaching Sahara: the threat to the West African colonies', *Geographical Journal* 85: 506–24.

Stebbing, E.P., 1937a, *The Forests of West Africa and the Sahara: A Study of Modern Conditions*. Chambers, London.

Stebbing, E.P., 1937b, 'The threat of the Sahara', *Journal of the Royal African Society*, Extra Supplement, May, pp. 3–35.

Stebbing, E.P., 1938, 'The man-made desert in Africa: erosion and drought', *Journal of the Royal African Society*, Supplement, January, pp. 3–40.

Stocking, M.A., 1972, 'Planting pattern and erosion on a cotton crop', *Rhodesia Science News* 6: 231–2, 236.

Stocking, M.A., 1984, 'Rates for erosion and sediment yield in the African environment', in *Challenges in African Hydrology and Water Resources*, Institute of Hydrology (IOH) Publication No.144, Wallingford, pp. 285–93.

Stocking, M.A., 1986, 'The cost of soil erosion in Zimbabwe in terms of the loss of three major nutrients', *Working Paper* 3, Soil Conservation Programme, Land and Water Development Division, Food and Agriculture Organization (FAO), Rome.

Stocking, M.A., 1992, 'Land degradation and rehabilitation: research in Africa 1980–1990: retrospect and prospect', *Drylands Networks Programme Issues Paper* 24, International Institute for Environment and Development, London.

Stocking, M.A., 1994, 'Assessing vegetative cover and management effects', in R. Lal (ed.), *Soil Erosion Research Methods*, 2nd edition, Soil and Water Conservation Society, Ankeny, IA.

Stocking, M.A. and Elwell, H.A., 1973, 'Prediction of subtropical storm soil losses from field plot studies', *Agricultural Meteorology* 12: 193–201.

Stoddart, L.A. and Smith, A.D., 1943, *Range Management*. McGraw Hill, New York.

Sutton, J.E.G., 1984, 'Irrigation and soil conservation in African agricultural history: with a reconsideration of the Inyanga terracing (Zimbabwe) and Engaruka irrigation works (Tanzania)', *Journal of African History* 25: 25–41.

Sutton, J.E.G., 1989, 'Towards a history of cultivating the field', *Azania* 24: 98–112.

Swanson, T.M. and Barbier, E.B. (eds), 1992, *Economics for the Wilds: Wildlife, Wildlands, Diversity and Development*. Earthscan, London.

Swift, J.J., 1988, *Major Issues in Pastoral Development with Special Emphasis on Selected African Countries*. Food and Agriculture Organization (FAO), Rome.

Taylor, C.C., 1935, 'Agriculture in Southern Africa', *Technical Bulletin* 466, US Department of Agriculture, Washington, DC.

Tegene, B., 1992, 'Erosion: its effects on properties and productivity of Eutric Nitosols in Gununo area, southern Ethiopia, and some techniques of its control', *African Studies Series* A9, Geographica Bernensia, Institute of Geography, University of Berne, Switzerland.

Theisen, R.J. and Marasha, J., 1974, 'Livestock in the Que Que Tribal Trust land: an ecological study of the importance of livestock development', mimeo. Agritex, Gweru, Tanzania.

Thies, E., 1993, 'Etude phytosociologique. Rapport final de la recherche d'accompagnement', Programme d'Aménagement des Bassins Versants Haute Guinée, Projet Kan 2, Conakry.

Thompson, H., 1910, *Gold Coast: Report on Forests*. Colonial Reports - Miscellaneous No. 66. HMSO, London.

Thompson, H., 1911, 'The forests of Southern Nigeria', *Journal of the African Society* 10 (38): 120–45.

Thompson, M., 1993, 'Good science for public policy', *Journal of International Development* 5 (6): 669–79.

Thompson, M., Warburton, M. and Hatley, T., 1986, *Uncertainty on a Himalayan Scale*. Ethnographica, London .

Throup, D.W., 1987, *Economic and Social Origins of Mau Mau, 1945–53*. James Currey, London.

Tidmarsh, C.E., 1952, 'Veld Management in the Karoo', Reprint No.4, Grootfontein College of Agriculture, Government Printer, Pretoria. From *Farming in South Africa*, January.

Tiffen, M., 1976, *The Enterprising Peasant: A Study of Economic Development in Gombe Emirate, North Eastern State, Nigeria*. HMSO, London.

Tiffen, M., 1987, 'Dethroning the internal rate of return: the evidence from irrigation projects', *Development Policy Review* 5 (4): 61–77.

Tiffen, M., 1990, 'Socio-economic parameters in designing irrigation schemes for smallholders: Nyanyadzi case study. Report IV: summary and conclusions'. HR/ODI/Agritex, Hydraulics Research, Wallingford.

Tiffen, M., Mortimore, M. and Gichuki, F., 1994, *More People, Less Erosion: Environmental Recovery in Kenya*. Wiley, Chichester .

UN, 1992, *Agenda 21: the United Nations Plan of Action from Rio*. United Nations, New York.

UNCOD, 1977a, *Round-Up, Plan of Action and Resolutions*, UN Conference on Desertification, Nairobi, 29 August–9 September. United Nations, New York.

UNCOD, 1977b, *Transnational Projects: Description and Status of Feasibility Studies*. A/CONF.74/3/Add.1. UN Conference on Desertification, Nairobi.

UNCOD, 1977c, *Sahel Green Belt Transnational Project*. A/CONF.74/29. UN Conference on Desertification, Nairobi.

UNCOD, 1977d, *Transnational Green Belt in North Africa*. A/CONF.74/25. UN Conference on Desertification, Nairobi.

UNEP, 1984, *General Assessment of Progress in the Implementation of the Plan of Action to Combat Desertification 1978–1984: Report of the Executive Director*, Governing Council, Twelfth Session, UNEP/GC.12/9. UN Environment Programme, Nairobi.

Union of South Africa, 1923, *Final Report of the Drought Investigation Commission*, Government Printer, Pretoria, U.G.49.

Union of South Africa, 1951, *Report of the Desert Encroachment Committee*, Government Printer, Pretoria, U.G.59.

Upton, M., 1987, *African Farm Management*. Cambridge University Press, Cambridge.

USAID, 1972, *Desert Encroachment on Arable Lands: Significance, Causes, and Control*. Office of Science and Technology, USAID, Washington, DC.

Vis, M., 1986, 'Interception, drop size distribution and rainfall kinetic energy in four Colombian forest ecosystems', *Earth Surface Processes and Landforms*, 1: 235–47.

Walker, B., 1985, 'Structure and function of savannas: an overview', in J. Mott and J. Tothill (eds), *Management of the World's Savannas*, Australian Academy of Sciences, Canberra, pp. 83–91.

Walker, B. and Noy-Meir, I., 1982, 'Aspects of the stability and resilience of savanna ecosystems', in B.J. Huntley and B.H. Walker (eds), *Ecology of Tropical Savannas*. Springen Verlag, Berlin.

Walker, B., Norton, G., Barlow, N., Conway, G., Birley, M. and Comins, H., 1978,

'A procedure for multidisciplinary ecosystem research with reference to the South African Savanna Ecosystem Project', *Journal of Applied Ecology* 15: 481–502.

Wallace, T., 1980, 'Agricultural projects and land in Northern Nigeria', *Review of African Political Economy* 17: 59–70.

Waller, R., 1976, 'The Maasai and the British 1895–1905', *Journal of African History* 17: 529–53.

Waller, R., 1988, '*Emutai*: crisis and response in Maasailand 1883–1902', in D.H. Johnson and D.M. Anderson (eds), *The Ecology of Survival: Case Studies from Northeast African History,* Lester Cook Academic Publishing, London.

Waller, R., 1993, 'Acceptees and aliens: Kikuyu settlement in Maasailand', in R. Waller and T. Spear (eds), *Being Maasai,* James Currey, London.

Warren, A. and Agnew, C.T., 1988, 'An assessment of desertification and land degradation in arid and semi-arid areas', *Drylands Papers* 2. International Institute of Environment and Development, London.

Warren, A. and Khogali, M., 1992, *Assessment of Desertification and Drought in the Sudano-Sahelian Region, 1985–1991.* UN Sudano-Sahelian Office, New York.

Warren, A. and Maizels, J., 1977, 'Ecological change and desertification', Background Document for the UN Conference on Desertification, Nairobi, 29 August to 9 September.

Warren, D.M., Slikkerveer, L.J. and Brokensha, D. (eds), 1995, *The Cultural Dimension of Development: Indigenous Knowledge Systems.* Intermediate Technology Publications, London.

Watson, R.M., 1991, 'Mkomazi - restoring Africa', *Swara* 14: 14–16.

Watson, R.M., Parker, I.S.C. and Allan, T., 1969, 'A census of elephant and other large mammals in the Mkomazi region of northern Tanzania and Southern Kenya', *East African Wildlife Journal* 7: 11–26.

Watt, M., 1913, 'The dangers and prevention of soil erosion', *Rhodesian Agricultural Journal* 10: 5.

Watts, M., 1983, *Silent Violence: Food, Famine and Peasantry in Northern Nigeria.* University of California Press, Berkeley, CA.

WCED, 1987, *Our Common Future:* Report of the World Commission on Environment and Development. Oxford University Press, Oxford.

West, O., 1948, 'Report on the grazing condition of Belingwe reserve', mimeo, Matopos Research Station.

Western, D., 1982a, 'Patterns of depletion in a Kenya rhino population and the conservation implications', *Biological Conservation* 24: 147–56.

Western, D., 1982b, 'Amboseli National Park: enlisting landowners to conserve migratory wildlife', *Ambio* 11: 302–8.

Western, D., 1994, 'Ecosystem conservation and rural development: the case of Amboseli', in D. Western and R.M. Wright (eds), *Natural Connections: Perspectives in Community-Based Conservation.* Washington, DC: Island Press.

Westoby, M., Walker, B. and Noy-Meir, I., 1989, 'Opportunistic management of rangelands not at equilibrium', *Journal of Range Management* 42: 266–74.

Whitmore, T.C., 1990, *An Introduction to Tropical Rain Forests.* Clarendon Press, Oxford.

Wilmsen, E., 1989, *Land Filled with Flies.* University of Chicago Press, Chicago, IL.

Wilson F. and Ramphele, M., 1989, *Uprooting Poverty: The South African Challenge.* David Philip, Cape Town.

Wilson, K., 1990, 'Ecological dynamics and human welfare: a case study of population, health and nutrition in Zimbabwe', unpublished PhD thesis, University of London.

Wood, G. (ed.), 1985, *Labelling in Development Policy.* Sage: London.

Worster, D., 1977, *Nature's Economy.* Cambridge University Press, Cambridge.

Worster, D., 1979, *Dust Bowl: The Southern Plains in the 1930s.* Oxford University Press, New York.

Worster, D., 1985, *Rivers of Empire: Water, Aridity and the Growth of the American West.* Oxford University Press, New York.

Worster, D., 1990a, 'The ecology of order and chaos', *Environmental History Review* 14 (1–2): 1–18.

Worster, D., 1990b, 'Seeing beyond culture', in 'A roundtable: environmental history', *Journal of American History* 76: 1078–106.

Wynne, B., 1992a, 'Risk and social learning: reification to engagement', in S. Krimsky and D. Golding (eds), *Social Theories of Risk.* Praeger, Westport, CT, pp. 274–97.

Wynne, B., 1992b, 'Uncertainty and environmental learning: reconceiving science and policy in the preventive paradigm', *Global Environmental Change* 2 (2):111–27.

Yair, A. and Lavee, H., 1976, 'Runoff generative process and runoff yield from arid talus-mantled slopes,' *Earth Surface Processes* 1: 235–47.

Yeraswork, A., 1988, 'Impact and sustainability of activities of rehabilitation of forest, grazing and agricultural lands supported by World Food Programme Project 2488', report to WFP and to the Natural Resources Main Department, Ministry of Agriculture, Addis Ababa.

Zimbabwe Government, 1986, *First Five Year Development Plan,* 1986–1990, Harare.

Zimbabwe Government, 1987, *The National Conservation Strategy: Zimbabwe's Road to Survival.* Natural Resources Board, Harare.

Zimbabwe Government, 1992, *Livestock Policy.* Ministry of Lands, Agriculture and Rural Resettlement, Harare.

INDEX